Fashion, Disability, and Co-design

FASHION, DISABILITY, AND CO-DESIGN

A Human-Centered Design Approach

GRACE JUN

BLOOMSBURY VISUAL ARTS
LONDON • NEW YORK • OXFORD • NEW DELHI • SYDNEY

BLOOMSBURY VISUAL ARTS
Bloomsbury Publishing Plc

50 Bedford Square	1385 Broadway	29 Earlsfort Terrace
London	New York	Dublin 2
WC1B 3DP	NY 10018	Ireland
UK	USA	

**BLOOMSBURY, BLOOMSBURY VISUAL ARTS and the Diana logo
are trademarks of Bloomsbury Publishing Plc**

First published in Great Britain 2024

Copyright © Grace Jun, 2024

Grace Jun has asserted her right under the Copyright, Designs
and Patents Act, 1988, to be identified as Author of this work.

For legal purposes the Acknowledgments on pp. 204–205 constitute an extension of this copyright page.

Cover design: Grace Jun, representing the three research domains: fashion, disability, and codesign.
Cover image © Vince Cavataio/Getty Images
Page ii image: Image courtesy of Zappos IP LLC.

Every effort has been made to contact copyright holders of photographs featured in this book.
Submissions regarding any inaccurate or missing acknowledgments can be made to the publisher.

All rights reserved. No part of this publication may be reproduced or transmitted
in any form or by any means, electronic or mechanical, including photocopying,
recording, or any information storage or retrieval system, without prior permission
in writing from the publishers.

Bloomsbury Publishing Plc does not have any control over, or responsibility for,
any third-party websites referred to or in this book. All internet addresses given in this book
were correct at the time of going to press. The author and publisher regret any inconvenience
caused if addresses have changed or sites have ceased to exist, but can accept
no responsibility for any such changes.

A catalogue record for this book is available from the British Library.

Library of Congress Cataloging-in-Publication Data
Names: Jun, Grace, author.
Title: Fashion, disability and co-design : a human-centered design approach
/ Grace Jun.
Description: London ; New York : Bloomsbury Visual Arts, [2024] | Includes
bibliographical references and index.
Identifiers: LCCN 2023021077 | ISBN 9781350299542 (paperback) | ISBN
9781350336940 (hardback) | ISBN 9781350299559 (pdf) | ISBN 9781350299566
(epub)
Subjects: LCSH: People with disabilities--Clothing. | Fashion design.
Classification: LCC TT648 .J96 2024 | DDC 746.9/2087--dc23/eng/20230602
LC record available at https://lccn.loc.gov/2023021077

ISBN:	HB:	978-1-3503-3694-0
	PB:	978-1-3502-9954-2
	ePDF:	978-1-3502-9955-9
	eBook:	978-1-3502-9956-6

Typeset by Typo•glyphix, Burton-on-Trent, DE14 3HE
Printed and bound in India

To find out more about our authors and books visit www.bloomsbury.com and sign up for our newsletters.

For Myungsook, Jae Whun, and Joung Hee.

Contents

viii **Preface**
Grace Jun, Assistant Professor of Graphic Design at the University of Georgia

x **Foreword**
Sara Hendren, Associate Professor of Arts, Humanities, and Design, Olin College of Engineering

xii **About the Author**

xiii **List of Abbreviations**

1. Introduction
2 **The Value of Collaboration in Fashion Design**
2 **Adaptive Fashion**
8 Characteristics of Adaptive Fashion Designs
11 Dressing Situations in Adaptive Fashion
14 Dressing for Social Integration
16 Dressing for Independent Living
18 Dressing for Interdependent Living
20 Dressing Aids and Assistive Tools

2. Theories, Models, and Methodologies
27 **Accessible Features in Existing Fashion Designs**
30 **Principles**
30 Inclusive Design
31 Fashion Design
33 **Processes, Models, and Methods**
33 Processes
33 Co-design
37 Models
37 FEA (Function, Environment, and Aesthetics)
39 MOHO (Model of Human Occupation)
40 HAAT (Human Activity Assistive Technology)
42 The Social Model of Disability
42 Method
42 UCD (User-Centered Design)
44 Human Factors

3. Approaches and Techniques
48 **Adaptive Fashion Framework**
49 Overview
50 Guideline
51 **Application of Framework**
53 Project Objectives
53 Establishing a Common Form of Communication
54 Making an Objective
56 **Research**
56 Observing Dressing Behaviors
56 Creating a User Journey Map
56 Recognizing Existing Clothing Challenges
58 **Design Requirements Method**
64 Define the Problem
65 Design Affordances
68 **Iterative Prototyping**
70 Materials
71 Textile Considerations
72 Body Measurements
72 **Techniques**
72 Clothing Closures and Construction Techniques
75 Alteration Instructions
87 **Conclusion and Design Evaluation**
87 Inquiry Guide
88 Design Rubric

4. Case Studies, Stories, and Interviews

- 94 **Interview: Inclusive Representation— Christina Mallon**
- 97 Unparalleled
- 101 Midi-Rox
- 108 The Little Black Bag
- 114 **Interview: Understanding the Use of Materials—Angela Domsitz Jabara**
- 119 Trans-Skirt
- 125 qXgo
- 133 Swipe
- 137 **Interview: Human Factors and Occupational Therapy— Michael Tranquilli**
- 141 Zipback Jacket
- 148 Ease
- 153 Modiste
- 158 Revolve
- 165 **Interview: Interactive Garments and Textiles—Dr. Jeanne Tan**
- 168 Versa Vest
- 173 Avisly
- 179 LIULID
- 185 Warmed Bomber

5. Conclusion

- 194 Increasing Visibility
- 197 Scaling Adaptive Fashion

200 **Afterword**
Steven Faerm, Associate Professor of Fashion at Parsons School of Design, The New School.

204 **Acknowledgments**

206 **Further Resources**

213 **Further Reading**

216 **Glossary**

217 **Index**

Preface

Grace Jun, Assistant Professor of Graphic Design, University of Georgia

Clothing is one of the oldest and most necessary forms of self-expression, but it's still not widely designed for one of the most marginalized groups in the world—PWD (people with disabilities). This is a problem for several reasons, one being that it can negatively impact someone's decision to engage in their community, speak up at work, or attend social gatherings that expect a particular dress code. While some big brands and retailers in the United States and elsewhere, such as Samsung, Target, Zappos, and Tommy Hilfiger, have introduced apparel designs for disability, also known as adaptive fashion, the process of designing with PWD is overwhelmingly absent (Lieber 2019). Adaptive fashion design is a complex operation that's influenced by three things: the product's objective, the designer's approach, and the product's intended user—who is ironically often excluded from the entire process. This reality highlights that inclusive design needs to be much more than simply creating something with PWD in mind, but rather an approach involving PWD in the design process. PWD are skilled at modifying and adapting inaccessible products, demonstrating that design processes can—and should be—approached from various viewpoints.

This book presents the complexities of fashion design processes when they are inclusive of disability needs. It presents methodologies, approaches, co-designed projects, iterative design processes, challenges collaborating with interdisciplinary groups, and how disability influences design experimentation. An example of such an approach is the relationship between functional clothing design and UCD (user-centered design) practice. Apparel design scholars have often used UCD to ensure a user's expectations are met, thereby promoting a solution that is inclusive of PWD (Watkins and Dunne 2015). User experiences, or lived experiences of PWD, are paramount in designing functional products that don't look like medical equipment (Lupton 2014). Still, PWD aren't always at the center of the design process, and the importance of aesthetics is often overlooked, even though this directly affects whether a product is purchased at all and eventually abandoned (The American Occupational Therapy Association 2020).

Lately, there's been a growing demand for fashion design curricula and coursework that's more inclusive of traditionally marginalized audiences—namely, curricula that teach students how to navigate the complexities of designing for consumers' disabilities and aging needs. In response to this need, several leading institutions are spearheading notable academic change. For instance, Parsons School of Fashion offered courses that collaborated with the Special Olympics and the disability-driven nonprofit Open Style Lab™ (Farra 2017; Feitelberg 2021), and students from North Carolina State's Wilson College of Textiles collaborated with the North Carolina Spinal Cord Injury Association to create adaptive fashions that drew on art inspiration. Drexel University's Master's program in Fashion Design offers adaptive fashion and inclusive design as one of the many studies within the program. Additionally, the Department of Human Centered Design at the College of Human

Ecology and the Master's program in Fiber Science and Apparel Design at Cornell University have long examined the relationship between textiles and functional garments applicable to disability needs. Like these institutional endeavors, designs inclusive of disability are emergent in publishing. Books like *Mismatch: How Inclusion Shapes Design* by Kat Holmes and *What Can a Body Do?* by Sara Hendren are influencing design attitudes and encouraging studies that examine the relationship between human experiences and products. The intersectionality of disability studies, communication design, human factors, and computer science is increasing in higher education.

Despite numerous studies on the relationship between design and disability emerging, particularly focusing on the built environment, there are few resources that show designers how to successfully apply co-design processes inclusive of disability to their fashion practice. Throughout these pages, you'll learn about unique challenges from various fashion projects that employed inclusive design theory, as well as one-on-one interviews with the clients and co-creators themselves describing the collaborative design process. Examples featured in this book include perspectives of PWD as designers and creative approaches examined by fashion design students. It's important to note that these stories don't feature cases working with every type of disability. Instead, they illustrate and explain how to apply thoughtful and collaborative fashion design that can be tailored to address any user's specific wants and needs. The methods and approaches to making fashion more accessible are ultimately framed with three guiding questions:

1. What role does clothing play for PWD?
2. How is a design process inclusive of disability explored?
3. How do personal experiences shape design outcomes?

People are increasingly questioning the attributes of what has historically been viewed as the norm in fashion, leading to a growing interest in meeting the needs of marginalized individuals. Diversity, inclusion, and equity are being embraced in everyday conversations, and it's important to extend that to disability. To better understand topics about disability inclusion, annotations for terminology like person-first language (e.g., PWD), identity-first language (e.g., disabled people), universal design, and inclusive design are provided as resources toward the end of this book. In addition, a list of adaptive designs, companies, and brands is also provided as a reference. The conversation points, a list of questions at the end of chapters 1 and 2, are specially aimed at supporting educators. These questions support student problem-solving and critical inquiry.

Moreover, design in this book is about the human experience. The human experience is central to critiquing how fashion design is practiced. How do people experience clothing as their bodies change? How do you measure an equitable design experience? Wearable designs, clothing that is comfortable, convenient, and considers well-being, provide a habitat for various bodies. How do design elements of clothing, such as material and fit, directly affect quality of life? In the following chapters, you'll discover why disability inclusion and collaboration are essential to the process of design fashion.

Foreword

Sara Hendren, Associate Professor of Arts, Humanities, and Design, Olin College of Engineering, USA

In the opening paragraphs of her Massey Lectures, broadcast in 1989 on Canadian public radio, physicist Ursula Franklin described her view of technology: "Technology is not the sum of the artifacts ... Technology is a system ... Technology involves organization, procedures, symbols, new words, equations, and, most of all, a mindset." (Franklin 1990).

Her celebrated remarks, now collected in book form, lay out a mindset for making things. Franklin called for the broad pursuit of "holistic" rather than "prescriptive" technologies: *holistic* design that proceeds from human and humane values, with its contexts of use deeply connected to human care, planning, and distribution. Our tools, instruments, devices, artifacts—and surely, too, our many rugged and beautiful articles of clothing—should not overly *prescribe* a narrow and homogenous way for working with them, she writes, at least not to the overwhelming degree wrought by mass manufacturing (Franklin 1990). So many of the tools we use are narrow and rigid in the tasks they make possible, such as a ballpoint pen, shoelace, or keyboard. While prescriptive objects "are often exceedingly effective and efficient," Franklin writes, "they come with an enormous social mortgage. The mortgage means that we live in a culture of compliance, that we are ever more conditioned to accept orthodoxy as normal, and to accept that there is only one way of doing 'it.'" (1990).

What would Franklin have made of the convivial riot of inventive gear that Grace Jun has assembled in this text? What an incredible treasure trove of fashion lies in these pages. An elegant coat for donning hands-free, dresses made to drape beautifully on wheelchairs, one-handed zippers, and magnetized button holes—plus clever adaptive loops, gussets, add-on pockets, and action pleats—this book utterly subverts any prescriptive idea of normal, whether among bodies or in the fashions made for them. Jun invites readers to take an open and holistic approach to design with and for disabled people, turning her expert eye on all the ways the world of fashion might loosen its seams a little, be edited, reassembled, or reconfigured in the pursuit of a more flexible idea of the body in the designed world.

So many everyday objects suffer from inevitability syndrome: the idea that the way things are is the way things must be and that the best humans can do is to fit their bodies and tasks to the tools at hand. But good design always proves otherwise. A finely honed power of attention will suggest new and flexible reuse, clever recombination, or fully redesigned gear that takes a holistic system to make real: relational insight, prototyping, iteration, and strong craft practices.

Part principled framework, part case-study collection, and part practical how-to guide, Jun provides an ideal text for design or textile studies researchers and beginner or experienced designers taking up systems-led, human-centered fabrication approaches to adaptive fashion. There are guidelines for understanding disability with proper ethics, close attention to human needs and wishes, and detailed exemplar designs with interview-style origin stories of how a coat or pair of pants came to be. With people at the beginning, middle, and end, and with attention to needs as well as wishes, pursuing pragmatism and beauty, this book is an open and holistic design system at its best.

In Grace Jun's hands, adaptive fashion—as an accessible ideal and as a way of working—provides an invitation both to thinkers and makers. "Design links theory and practice," Jun writes, and "elegantly manages the complexity of open-ended problems" (2024), like the search for adaptive wearable design. *Fashion, Disability, and Co-design* does this same elegant management, reframing big ideas while also instructing at the small detail level. In doing so, Jun circumvents the supposed opposition between universal, scalable design and bespoke customization. Instead, so many of the ideas in this text inspire new questions about design for small-, medium-, and large-scale production: Might a magnetized closure be used for a whole new line of jackets? Might an extra pocket come in a kit of parts, attachable to a shirt or dress at a personalized location? There are ideas for hacking and ideas for interrupting large manufacturing processes. Jun's text is flexible and open-ended, pointing readers "less toward finding 'answers' and more toward 'methods leading to answers.'" (2024).

In the long legacy of postwar rehabilitation engineering—with its prosthetic parts, high-tech tools, and futuristic fantasies—there is a wider, deeper material culture of disability all around us. Adaptive fashion is just one of the many places where design meets disability in a much more everyday context. But this very quotidian modesty is what makes adaptive design so radical. Remaking the world in tangible, affordable, often invisible ways makes moving in and through the world possible. This book, celebrating Grace Jun's work and the work of so many other visionary designers, is about the daily task of getting dressed. And getting dressed means getting out the door and into public life.

About the Author

Grace Jun is assistant professor of graphic design at the University of Georgia (UGA) researching design processes inclusive of disability, such as accessible interaction design, co-design, and inclusive fashion. Her professional experience has included different positions in industry, academia, and entrepreneurial business. This has influenced her research and teaching to be interdisciplinary and collaborative. Grace is a founding member of Open Style Lab™, a Smithsonian National award-winning nonprofit organization that aims to make style accessible for all people regardless of cognitive or physical disabilities. She is also a proud alumnus of both Parsons School of Design and Rhode Island School of Design and was award Emerging Designer of the Year. Grace continues to serve on jury committees and organizations that advance the arts & design. She resides in Atlanta, Georgia with her husband, Dr. Gregory D. Lee.

Abbreviations

AATCC	American Association of Textile Chemists and Colorists
ADA	Americans with Disabilities Act
ADLs	activities for daily living
AFO	ankle-foot orthoses
AI	artificial intelligence
AiDLab	Laboratory for Artificial Intelligence in Design
ALS	amyotrophic lateral sclerosis
APA	American Psychological Association
ASD	autism spectrum disorder
AT	assistive technology
CAD	computer-aided design
COPD	chronic obstructive pulmonary disease
CPG	consumer packing goods
DE&I	diversity, equity, and inclusion
DIY	do-it-yourself
DVT	deep vein thrombosis
ePTFE	expanded polytetrafluoroethylene
EVA	ethylene-vinyl acetate
FEA	Function, Environment, and Aesthetics
FIM	Functional Independence Measure
FIT	Fashion Institute of Technology
HAAT	Human Activity Assistive Technology
HKPOLYU	Hong Kong Polytechnic University
IADLs	instrumental activities of daily living
MDA	Muscular Dystrophy Association
MOHO	Model of Human Occupation
MS	multiple sclerosis
OPHI-II	Occupational Performance History Interview-II
OCD	obsessive-compulsive disorder
OT	occupational therapist
PE	polyester
PLA	polylactic acid
POF	polymeric optical fiber
PVA	polyvinyl alcohol
PWD	people with disabilities
RA	rheumatoid arthritis
SCI/D	spinal cord injury or damage
SMA	spinal muscular atrophy
TPU	thermoplastic polyurethane
UCD	user-centered design

Chapter 1
Introduction

The Value of Collaboration in Fashion Design

While clothing is ubiquitous, the process of fashion design is complex. Designers are expected to utilize a range of skills, including visualization, construction techniques, critical inquiry, material knowledge, and holistic systems management. However, the method in which they combine these skills and the order in which they do so is usually enigmatic. There is no one way to practice design but rather multiple ways to approach the body, dress, and human experience. Sometimes, designers discover methods or change their approach while creating a design. Creative practices like fashion, therefore, ask designers to be flexible in their approaches and thinking. Fashion is not static. As Loschek writes, fashion "is the interface between creation and social communication, between form and medium" (Loschek 2009). Consequently, fashion design processes are in continuous flux with an ever-evolving appearance.

Fashion design thinking manifests the material. It is a preparation for creating something tangible. What constitutes a successful **co-design** fashion experience is, therefore, intentionality—setting up conditions that foster great design work and identifying shared values. A designer's intention to collaborate with other stakeholders or co-designers (e.g., disabled people, tailors, or fashion merchandisers) at the beginning of a project is different from designing for people as an afterthought. When and how each co-designer participates in the process ultimately influences the final product. A participatory learning experience, where insights and values are exchanged between participants, is paramount to that process. Values are not visible but are experienced through interaction. Participants can learn about resourcefulness, sharing knowledge and best practices, and troubleshooting issues as they arise in the co-design process. Danielle Allen discusses that participatory readiness fosters civic engagement in our communities (Allen 2016). Similarly, social awareness and community engagement in fashion design will likely result in an inclusive process that engages with more people—particularly those often overlooked by and marginalized in society. Collaborative and inclusive processes are likely to develop valuable or beneficial experiences for more people.

Adaptive Fashion

Fashion has enormous potential to change society's attitudes and ways of thinking about bodies and clothing through **adaptive fashion**. Described as the "social by-product of the opposition of conformity and individualism and of unity and differentiation, in society," clothing mediates the self and the external environment (Simmel 1957). Clothing provides a unique way for people to express themselves. And although designs created for and by **PWD (people with disabilities)** have existed, these products are rarely depicted in the mainstream marketplace. Moreover, early US literature has focused primarily on functional details that satisfy the clothing needs and wants of PWD rather than the modes of personal expression (Dallas 1982; Lamb 2001; Freeman 1985). "Clothing for the handicapped," "functional," and "adaptive" are labels all used to describe clothing designed for PWD and other impairments (Kernaleguen 1978; Gupta 2011; Carroll 2015). Notable scholars like Anne Kernaleguen and Jane M. Lamb described functional clothing as rehabilitative tools but also design opportunities to alter ready-to-wear clothing. Additionally, academic researchers like Carroll, Gross, and Quinn began noting the deficit in the commercial availability of functional clothing (Carroll and Gross 2010; Quinn 1990). An "evolution from functional clothing to designs that focus more on the aesthetics and styling needs and wants of people living with disabilities" is presented by scholars McBee-Black and Ha-Brookshire (2020). This transition is evident with brands and retailers emphasizing the words "adaptive fashion." For example, *Vogue* notes adaptive fashion is in the spotlight and highlights brands such as Kohl's, Tommy Hilfiger, and JCPenney (Berlinger 2022).

Adaptive fashion is one clothing option available to PWD that integrates both functional (determined by the garment's construction and materiality) and social needs (influenced by culture and social trends). For example, a pant design can function as a cover that protects the body from harsh weather conditions based on the choice of fabric (The American Occupational Therapy Association 2020) and communicate personal style. This further validates that fashion design is dynamic—unfixed and evolving. Furthermore, adaptive fashion is adjacent to the assistive paradigm. For example, prosthetics and dressing devices both assist the body and impact how people dress. The assistive paradigm also includes adaptations and modifications. Similarly, in fashion, body modifications, alterations, and adaptations became especially notable in the seventeenth century in Europe because of corsets, bustles, girdles, and brassieres, along with other period garments that were molded by these distinctive understructures. Exhibitions like *Undressed: A Brief History of Underwear*, which ran from April 16, 2016 to March 12, 2017 at the Victoria and Albert Museum in London, examined how devices have shaped the human figure. Exemplary underwear pieces explore how lacing, straps, and stretch fabrics can alter natural body forms. The 2001 *Extreme Beauty: The Body Transformed* exhibition, which ran from December 6, 2001 to March 17, 2002 at The Metropolitan Museum of Art in New York, presented "fashion as the practice of some of the most extreme strategies to conform to shifting concepts of the physical ideal" (The Metropolitan Museum of Art, n.d.). Each section of the exhibition focused on specific parts of the body—neck, shoulders, bust, waist, hips, and feet—and how people deformed, manipulated, exaggerated, and constricted the body. The need to manipulate the body through dress is like some of the features offered in adaptive fashion. For example, the dressing process for PWD is impacted, changed, and assisted by devices or body adaptations as referenced in chapter 4, Trans-skirt.

Extensions of the body, or what is part of the body, are areas of study that influence adaptive fashion design. From voice-activated watches to lightweight electric wheelchairs, defining what **wearables** are is subjective to a person's identity and body. There are various conceptual thinking models found in **AT (assistive technology)** that can frame adaptive fashion design approaches. Conceptual thinking models, such as the HAAT (Human Activity Assistive Technology) model, have led to a preponderance of product outcomes for PWDs (Cook and Hussey 2002). These include wearable solutions attached or worn on the body such as braces, prosthetics, or other amplifications that explore similar dressing challenges found in adaptive fashion. For example, a person with lower limb amputation may need pants that prevent abrasion of a prosthetic leg or have a detachable section to be worn easily. Helen Cookman's patented trouser modification includes an elastic waistband, large pockets, and full-length side seam zippers for easy donning and doffing. The pant design has a belt that specifically provides support when using a restroom. This design already assumes the body must adapt to situations where restroom use is not easily accessible—a criteria not being met in available clothing. Most importantly, Cookman's design coincides with fashion trends during a "time where blue jeans were the thing to wear," notes Tracy Panek (pers. comm., 2022), Historian and Director at Levi Strauss & Co.

Archives. The 1970s witnessed a surge in consumer desire for denim, seen in flare and bootcut jeans. The images below provided by Levi's® demonstrate how the trouser design was not only functional but strategically fashionable.

After the Second World War, medical professionals like Howard Rusk, a prominent figure in rehabilitation medicine, recognized that rehabilitation solutions were important for all Americans and hired Cookman to explore the need of **functional fashion**. She worked at the Rusk research residency before collaborating with Levi's® in the 1970s to create Functional Fashions, a sample line created to help PWD dress independently by testing "function, utility, and fabric choice" (Cookman and Zimmerman 1961). As a designer with a hearing impairment, Cookman's work is a notable milestone in adaptive fashion clothing studies and the disability community in the United States. Her work was featured in the research project and exhibition *Functional Fashions*, which was displayed at the Milwaukee Art Museum through the spring of 2020. Similar images of her work

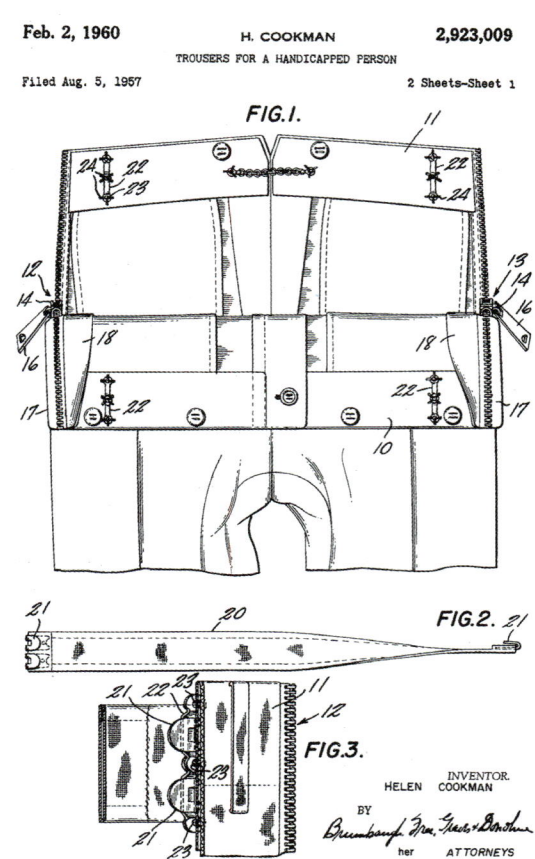

Figure 1.2 US patent of pant design by Helen Cookman.

Figure 1.1a–b Pant design by Helen Cookman and Levi's® under the Functional Fashions collaboration. Images from Levi Strauss & Co. Archives.

Fashion, Disability, and Co-design

provided by the Medical Archive and Special Collections at the Rusk Institution in New York University Langone Health featured clothing with shirt pleats, Velcro® fastenings, and extra layers of fabric underneath garments with sleeves to withstand the pressure from crutches—common designs today that can benefit everyone. Designers like Cookman demonstrate that PWD are skilled at modifying and designing products to be accessible, highlighting that design processes can—and should—be approached from various viewpoints.

Despite Cookman's designs, the environment and conditions during wartime were hardly conducive to the growth and visibility of adaptive fashion. Rehabilitation needs and technological enhancements in textiles dominated US products. Prosthetics were introduced as wearable solutions, but they were primarily functional innovations that were seen as ways to enhance or improve primary war veterans and integrate them into the American workforce (Williamson 2019). Early designs were missing the inclusive cues that fashion provides all Americans—the opportunity to fully express oneself through clothing and participate in society. Adaptive fashions were not readily seen in the market, nor were disabled designers more recognized. Disabled makers and fashion designers with disabilities can be traced in US patent records. Harry Spack, whose legs were injured in the Second World War, invented a pair of inflatable trunks for people with restricted mobility to swim or float more safely (Spack 1946). Patents like these trunks may provide records of early disabled designers and adaptive fashions. Disabled makers and research into disability and design patents can be referenced by the work of Kat Jun and Kate Annett-Hitchcock (see Further Reading).

Another contributing factor to the lack of stylish yet functional clothing choices for Americans with disabilities in the 1940s was **accessibility** to information. "You couldn't just buy these [Helen Cookman's jeans] in department stores. Postcards and newsletters in doctors' offices had to be requested, so it was niche," says Tracy Panek (pers. comm., 2022).

Early literature studies suggest nurses and educators provided instructions on dressing for and with a disability but not design solutions.

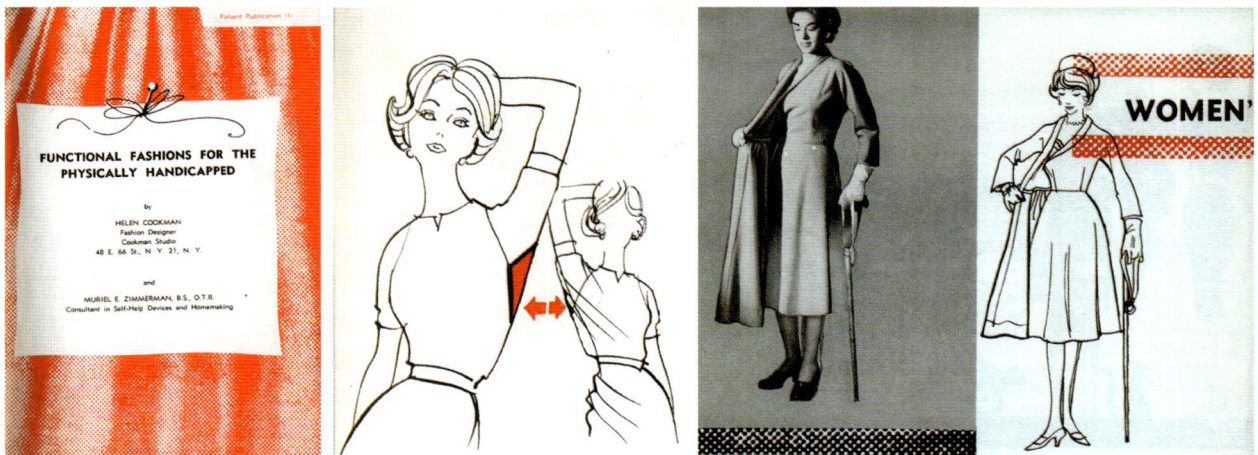

Figure 1.3a–c Images from *Functional Fashions for the Physically Handicapped*, including an underarm gusset and wrap dress design. Images courtesy of The Lillian and Clarence de la Chapelle Medical Archives at NYU.

Marion Kerr writes extensively on nursing responsibilities that include learning barriers and dressing instructions for people with cerebral palsy (Kerr 1946). Dressing solutions did not focus on inaccessible designs, but rather on how people could adapt existing designs to their disability through training and education. This further perpetuates social pressures and nonacceptance faced by disabled people to conform to specific standards. Edith Sagul reports the experience of one teacher who was disappointed by the appearances of her cognitively disabled students: "The children came to school in clothing which was held together with [a] contrasting color of thread, unmatched buttons, and safety pins serving for ties, buttons, and patches. Some ties hung like ropes and were sorely in need of cleaning" (Sagul 1943). A lack of adaptive clothing is a social concern because the students were ostracized and separated from others by their appearance.

Social pressures to conform to idealized dress codes were felt not only by students with disabilities but by all Americans during the Second World War. Citizens were expected to adhere to idealized American standards of cleanliness, health, and efficiency. It was a time when American fashion was also distinguishing itself from European designers with styles that were utilitarian and simple, as seen in designs by Vera Maxwell and Claire McCardell. With a ban on raw materials, fashions during this time were designed with an emphasis on frugality, conformity, and versatility in use. Social class differences were less visibly noticeable as the dress and style of all women became similar. All women, even stout women, were expected to wear things that upheld these expectations. Lane Bryant "realized the need for clothing that fit a larger figure," and the brand's early plus-size clothing offered choices of body diversity and conformity for stout women, yet disability was rarely considered (Keist 2017). By the end of the Second World War, fashion silhouettes changed dramatically, and the fashion industry flourished with important crossovers between film, theater, and fashion (Deihl 2018). Paris couture houses held influence in American postwar fashion, such as Dior's Corolle line, also called the New Look, an overt reaction to the restrictions of materials being banned or rationed. An example of Dior's New Look was the Chérie, spring/summer 1947 (available to view on The Met's Collection website), a skirt made of the full width of the fabric, creating a substantial bulk with pleats.

Figure 1.4 Levi's® announcement letter for functional jeans in 1975. Image from Levi Strauss & Co. Archives.

With rations eased after the Second World War, textile choices and fashion styles were changing. Oversized zoot suits and high demands for tight-fitting nylon stockings demonstrated materials and silhouettes of clothing were changing mainstream styles. Fashion was making unprecedented advancements in both aesthetics and textile technology, such as fabric dyes, synthetic textiles, and new and faster machinery, were all propelled forward. Nylon is described as "the glamour thread that makes stocking more sheer than silk" and is featured as the tough and versatile material innovation in the 1940s (Morrow 1948). Furthermore, American handcraft was seen as a tool to facilitate cultural and economic change during and after the war. Textiles were used in a variety of domestic and international situations, such as the American Red Cross using textiles to rehabilitate injured veterans (Troy 2014). Despite this, there was a missed opportunity where disability needs could have overlapped with textile and fashion advancements. Disabled people could have been more directly involved instead of being treated as a group of people who needed charitable service within fashion and textile processes. Designing with disability was still influenced by societal prejudices and recognizing diverse user needs was not as prominent until the 1980s with rise of **universal design**.

Figure 1.5 An afternoon suit jacket in natural shantung and corolla skirt in pleated wool from Dior's spring/summer 1947 Haute Couture collection, Corolle line. Photography by Grace Jun at the spring 2022 *La Galerie Dior Exhibition* in Paris. Courtesy of Grace Jun.

Introduction 7

Characteristics of Adaptive Fashion Designs

Like all apparel, adaptive fashion is a visual language that has its own style, form, and features. Over the last decade, it's grown in popularity with a few distinctive visual and tactile elements: silhouettes that allow easier donning and doffing (the process of wearing on and taking off a garment), fasteners that are strategically placed, sensory-friendly fabrics, and fittings tailored for the seated position. From Tommy Hilfiger's Tommy Adaptive collection (launched in 2018) to Heartist by Samsung C&T Fashion Group (launched in 2019), more brands are beginning to recognize the clothing needs of PWD, echoing the design solutions proposed earlier by Cookman.

Tommy Adaptive, one of the most recognizable American adaptive fashion lines, has garments that include magnetic and hook and loop closures that aid individuals with arthritis or limited mobility with their hands or fingers. Because the type and placement of fasteners affect a wearer's movements, the company offers such features in locations that provide ease of dressing. Polo shirts and athletic jackets are a few examples of the designs presented by Tommy Adaptive. An early advocate for PWD, Tommy Adaptive demonstrates how adaptive fashion can be implemented across various categories, such as athletic wear and

Figure 1.6 Adaptive shirt closures for men in *Functional Fashions for the Physically Handicapped*. Image courtesy of The Lillian and Clarence de la Chapelle Medical Archives at NYU.

Figure 1.7 A Velcro® and loop closure on Heartist's denim pants for men. Courtesy of Heartist.

business casual. In doing so, Tommy Adaptive integrates disability needs holistically through fashion branding.

Hard-to-reach areas, such as the back of a person, greatly benefit from having easier and more accessible clothing closures. Such designs are generally lacking from most brands, thus making Tommy Adaptive an early advocate for PWD in mainstream fashion. More so, the sport-like designs in Tommy Adaptive demonstrate how an adaptive line addresses similar dressing challenges that could apply to other apparel lines, such as athletic wear.

Hidden magnets, adjustable hems, larger pockets, one-handed zipper, and elastic closures not only make garments accessible to use for PWD but also better for athletes. For example, a long zipper along the pant side seam can provide running athletes great flexibility and comfort when putting on shoes.

Materials also play a large role in the fashions seen in sportswear. For example, performance materials can protect, provide comfort, and be sustainable, like the design process of Abby Gaskin and Mickey Chan. Abby met Mickey, a New York Special Olympics member, at Parsons School of Design in a fashion design course that collaborated with the Special Olympics. They collaborated to create designs that could better facilitate Mickey's skating movements by discussing ways clothing can protect against the wetness of ice and be easier to wear against skin sensitivities. Using a collaborative design process, they focused on finding sustainable materials to create outfits worn when ice-skating.

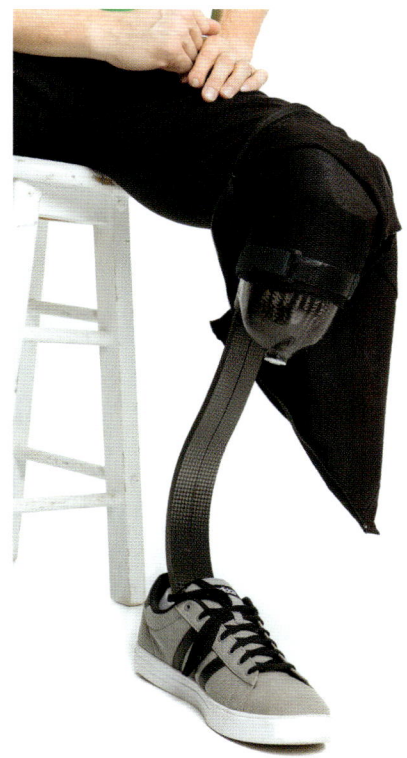

Figure 1.8a–b. No Limbits men's adaptive pants using zipper closures for wider access for leg or prosthetic. Images courtesy of Zappos IP LLC.

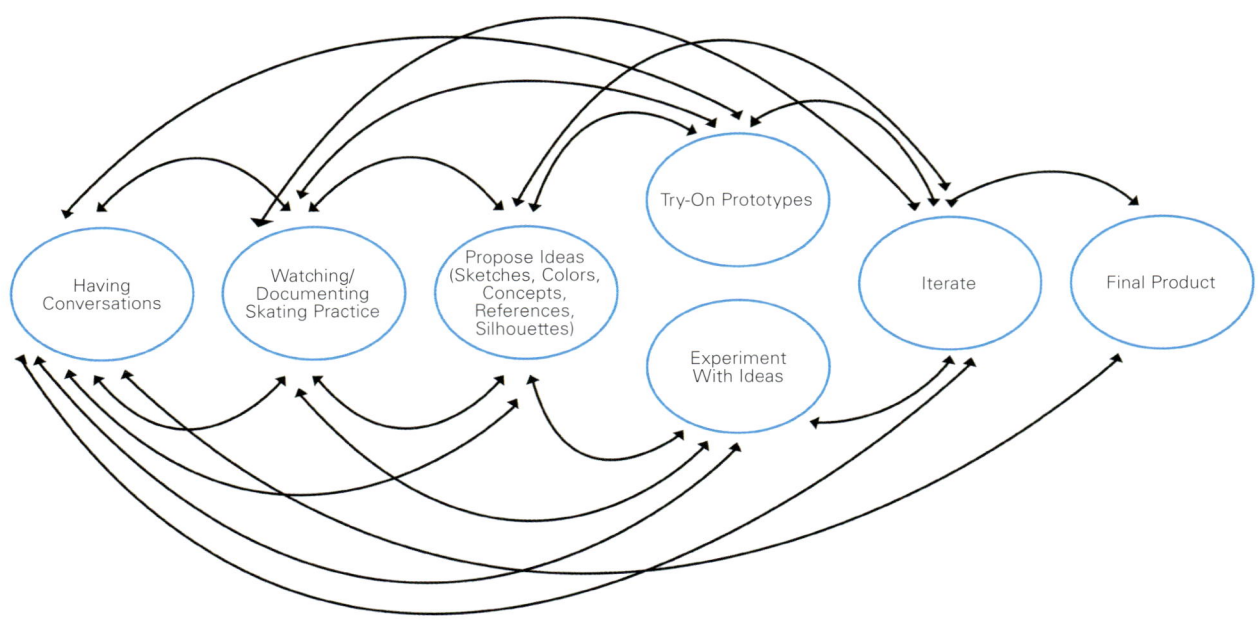

Figure 1.9 Co-design process sketches by Abby Gaskin and Mickey Chan. Image courtesy of Abby Gaskin.

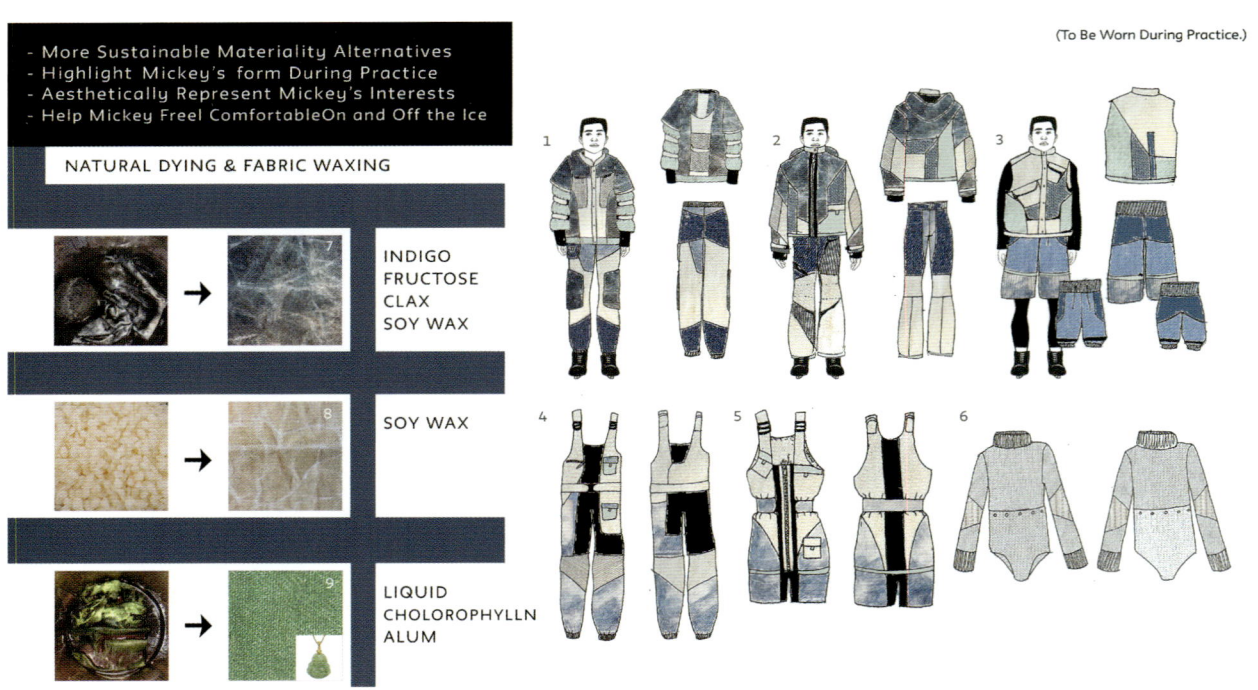

Figure 1.10 Material and garment sketches drawn by Abby Gaskin. Image courtesy of Abby Gaskin.

10   Fashion, Disability, and Co-design

"We experimented on cotton twill fabric using soy-based wax with natural dyes, which achieved a more water-resistant material for Mickey's practice outfits," says Abby. The results were a collection of various outfits using natural blue dyes to explore sustainable approaches toward adaptive sportswear.

Dressing Situations in Adaptive Fashion

Aging shows us that different degrees of ability are simply a part of living, which design has the power to support. Adaptive wear designer Ellen Fowles created clothing that supports wearers both during receipt of medical care and in everyday use, providing comfort and destigmatizing the experience of being a patient. Her design practice is achieved through intergenerational co-design and research in **ethnography** to create garments that people of all ages and abilities can enjoy. This is examined in a capsule collection created with and for her grandmother, Marian Fowles. The collection aimed to create a wardrobe that accommodated Marian's lifestyle at home and in the hospital.

Similarly, Wendy Wong, Founder of the Canadian brand June Adaptive, witnessed close family members with medical conditions such as arthritis, paralysis, dementia, and MS (multiple sclerosis) struggle to find clothing solutions to their mobility challenges. Named after Wendy's aunt, June, who lost mobility in her arms and legs due to an accident, June Adaptive aims to help PWD find easy-dressing contemporary fashions. All products aim to provide people with minimum fatigue to get dressed. Thoughtful designs like their adaptive shoe and side-zipper top help reduce or remove barriers associated with aging and disability.

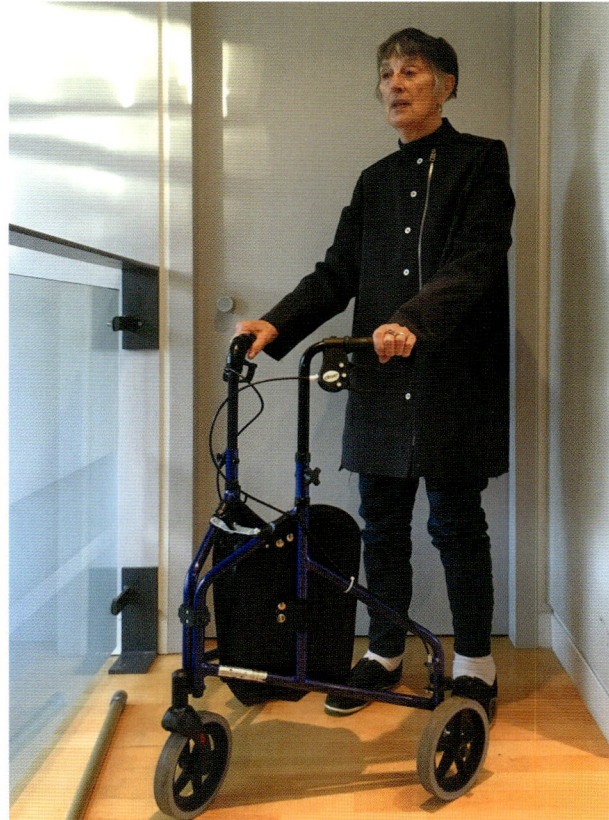

Figure 1.11 Marian Fowles wearing the Bib Shirt made from organic brushed cotton with a detachable water-resistant chest cover featuring a magnetic engineered zip closure, decorative corozo buttons, and a hidden handkerchief pocket. Image courtesy of Ellen Fowles.

Figure 1.12 Side-zipper access shoe design and grip socks showcased by founder Wendy Wong for June Adaptive. Courtesy of JuneAdaptive.com.

Figure 1.14 A purple top with a discrete zipper that increases the shoulder and neckline opening worn by model Chelsea for June Adaptive. Courtesy of JuneAdaptive.com.

Figure 1.13 Rear-zipper access shoe designs worn by model Steve for June Adaptive. Courtesy of JuneAdaptive.com.

Inclusive design, co-design, and participatory design are all practices that specifically respond to the human experience. Dressing situations for all ages and disabilities change throughout a person's lifetime. For example, sensory-friendly apparel can mean different things to different people. For the visually impaired or blind, soft textures and differentiating textures using embroidery or tactile labels are useful for identifying pieces. Similarly, the Two Blind Brothers also integrate such tactile features in their men's and women's clothing collection, as shown on their website, readily available for purchase with accessibility considerations for the blind community.

Two Blind Brothers

Apparel also provides comfort for people who experience skin irritation due to fabrics, detergents, and dyes, greatly benefiting individuals with autism. To support babies and children with sensory sensitivities, Target created Cat & Jack in 2016. The **adaptive designs** of the brand are based on the shared experiences of their children, children with special needs, and common customer requests, such as (1) removing tags and skin-irritating embellishments, (2) offering high-rise leggings that fit over diapers (The American Occupational Therapy Association 2020), and (3) featuring extra-soft, durable knits. The success of this company highlights three ways that dressing experiences can be contextualized: social integration, individual independence for PWD, and interdependent relationships.

Target Adaptive

Finally, adaptive fashion designs include accessories. From must-have handbags to everyday backpacks, accessories serve as both a statement of personal style and extra portability to carry more items close to the body. Iconic brands like JanSport feature an Adaptive Backpack and Adaptive Crossbody that can be worn or placed on strollers, wheelchairs, and other assistive devices.

JanSport Adaptive Backpack and Adaptive Crossbody

Similarly, FEELDOM is a company of designers and craftspeople in South Korea working with disability communities to create higher-quality wheelchair bags.

FEELDOM Life

Figure 1.15a–b A person reaching back to use a FEELDOM wheelchair bag's top pocket. The change modes from a wheelchair bag to a shoulder bag with a strap. Design by Adrianne Mascho and Mikyeong Kim. Courtesy of FEELDOM Co., Ltd.

Dressing for Social Integration

"My two requirements are that [the jacket] needs to be sleek and functional," says Quemuel Arroyo, Chief Accessibility Officer at the New York Metropolitan Transportation Authority (pers. comm.). Quemuel's desire for a stylish yet professional look is shared among many working Americans living in densely populated cities, which is also referenced in chapter 4, qXgo. Clothing can support or hinder the achievement of an individual's everyday activities and the fulfillment of social roles in the workplace, which is often overlooked for someone living with a disability. Workplace attire, such as required uniforms and unspoken dress codes, have been barriers for PWD to engage fully in society. For example, weddings and funeral attire are rarely designed for the seated body, which may exclude wheelchair users from participating in cultural milestones and other important events. The lack of appropriate clothing complicates achieving and maintaining employment for PWD. Fashions inclusive of disability are presented on a variety of scales, from medium-sized businesses to large conglomerates. Accessible workwear clothing has a substantial impact on quality of life, building relationships in the workplace, and employment, as seen in the collection Heartist.

Heartist is one of Samsung Construction & Trading Corporation Fashion Group's major corporate social responsibility brands. Samsung C&T Fashion Group realizes the value of designing the lives of everyone around the world through fashion with the vision of "We Design for Life." Heartist was designed in response to South Korea's disabled population, particularly wheelchair users. The Heartist brand focused on wheelchair disabilities by collaborating with fashion experts, rehabilitation medical treatment specialists and, most importantly, PWD at the Samsung Medical Center and the Disabled First Movement Headquarters, both located in Seoul. Heartist's first collection consisted of twenty-seven styles of jackets, blouses, T-shirts, pants, and skirts for the spring/summer season of 2019.

The collection showcased business casual clothing with both functionality and design that first started with the intent that the dressing experience for the workplace should be easier for the lifestyle of wheelchair users. The brand's key features included an elastic fabric applied to the back for jackets, increased crotch length, and belt ring reinforcement for pants to increase durability when pulled. Trousers worn by people using wheelchairs must be higher in the back and lower on the front waistline with a wider crotch area. The pants were

Figure 1.16 Two models wearing Heartist designs for men's and women's fashion. Courtesy of Heartist.

Figure 1.17 a–b Women's jeans with a seated design, including an elastic waistband, designed by Heartist. Courtesy of Heartist.

also designed in a silhouette that prevents excessive pressure and blood flow restriction. Finally, thick seams were avoided, especially around the parts of the body that remain seated.

Hallmarks of the brand's apparel include designs that avoid using excess fabric around the abdomen. While keeping the business jacket aesthetic, the jacket was designed to avoid bunching of fabric with a shorter front hem and to prevent garments from catching on wheels and other places. Additional ease in sleeves and shoulders was placed to respond to the rolling movement when using a manual wheelchair. Stretch fabric was chosen because it provides extra room for arm movement, as seen in figure 1.19.

Heartist Collection

PWD face barriers to employment, but clothing should not be one of those restrictions. Workwear garments indisputably are part of our professional identity, no matter where we work or what we do for a living. Workwear can serve as an entry point or barrier for many people to participate in social activities and obtain greater independence.

 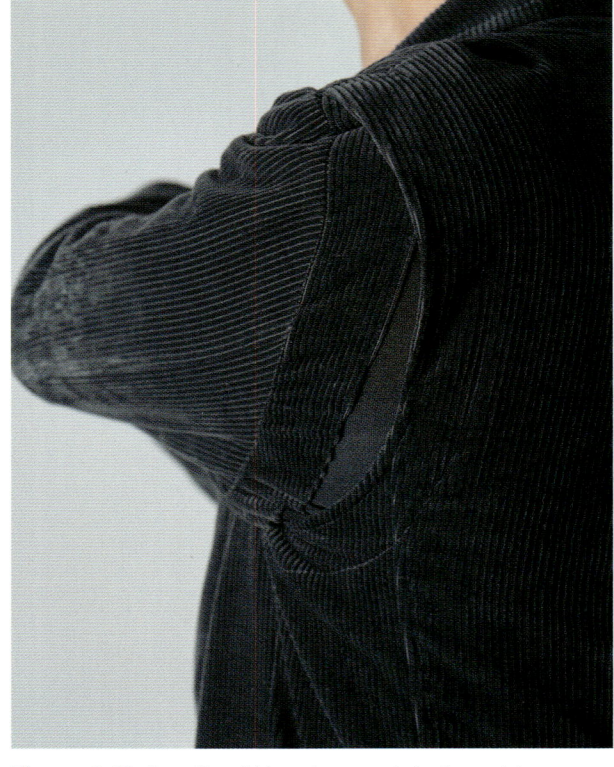

Figure 1.18 Men's jacket design by Heartist featuring a shorter hem and larger pockets.
Courtesy of Heartist.

Figure 1.19 Details of Heartist men's jacket with an elastic action pleat for a seated design.
Courtesy of Heartist.

Dressing for Independent Living

Adaptive fashion allows PWD to look stylish and dress independently, increasing their self-esteem, self-confidence, and overall quality of life. The idea that a person's attire can affect their thought processes is described as "enclothed cognition," based on research conducted by Adam and Galinsky (2012). The study explored the effect of attire on attention for task performances using lab coats (Adam and Galinsky 2012). The ability to dress independently can hinder or increase the self-confidence of a person recovering from injury. Singaporean designer Claudia Poh's Werable is an adaptive fashion collection that addresses **independent dressing**. Werable includes a multifunctional top that acts as both an arm sling and bolero and also pants that are designed to integrate catheters.

With a research process focused on relationships with people who have experienced a stroke, Claudia describes how co-creation was essential in her approach: "Before starting Werable, I was based in New York City, where I co-founded Cair Collective with Amy Chen. As a team, we made a collection enabling a hands-free dressing experience with Christina Mallon, who was diagnosed with ALS (amyotrophic lateral sclerosis). We learned that all these hacks and innovations can also help us discover new ways of dressing for all" (Poh, pers. comm., 2022).

The bolero-sling garment demonstrates the versatile nature of fashion and the co-creation process. In collaboration with Stroke Support Station, Singapore's first stroke rehabilitation center, Claudia was commissioned to design an arm sling "to look like a shawl." (Poh, pers. comm., 2022). The design evolved to also allow for extended use after recovery. The Stroke Support Station encouraged Claudia and stroke survivors to consider how fashion can be used as a tool for reintegration into work, life, and society.

Figure 1.20a–b Garment design from the Werable collection designed by Claudia Poh. Images courtesy of Werable.

Figure 1.21a–b Pant design with integrated catheter pocket designed by Claudia Poh. Images courtesy of Werable.

Dressing for Interdependent Living

The act of dressing does not always have to be a solo endeavor. While achieving independent dressing can be a goal for some people, there are opportunities where co-dressing can reexamine our relationships with clothing as an **interdependent dressing** experience. For example, models preparing for the frantic couture runway are often assisted by many people to get dressed. Similarly, older PWD living in rehabilitation centers need the assistance of attendees to do the same. Yet the context and situation of luxury couture versus rehabilitation dramatically change how people view and experience assistive dressing. This example is extreme in differences of environmental context and socioeconomic status, yet the action of getting dressed with assistance is shared. Dressing can become a complex experience for people who strongly rely on others for intimate daily activities, such as getting undressed when using a restroom. Designing adaptive fashions that consider assistance and co-dressing from the start of conception could provide easier and quicker dressing experiences for all abilities, as shown with the Warmed Bomber in chapter 4.

Figure 1.22 NOT Interlock coat worn by dancer and director Elena Vazintaris. Courtesy of NOT and Susanne Vogel.

Figure 1.23 NOT split top worn by model Muriel Favaro. Courtesy of NOT and Andrew Boyle.

Some fashion designs don't begin with disability needs yet still offer accessible solutions and creative inquiries about the relationship between body and material. NOT, a New York-based brand led by creative Jenny Lai, aims to make dressing a surprising, playful, and transformative experience. Featuring women's and men's ready-to-wear, as well as custom performance wear for musicians and dancers, NOT looks at how clothing functions and behaves in ways that don't impede performers, therefore giving them greater freedom of movement.

While NOT doesn't identify as a disability adaptive line, its models are diverse in gender and age, and many of its garments have accessible features—such as breathable fabrics that don't stick to the body and an omission of hardware closures.

Designs like Tiffany Hwang's Beloved collection invite people to consider relational and spatial awareness of bodies. "You can't assume people will feel comfortable without consent and giving that space," (Hwang, pers. comm., 2022) says Tiffany. An integral part of Tiffany's initial design inquiry was photography, which she used to observe bodily interactions among her friends.

Tiffany was able to see people's comfort levels through proximity—some would stand very close to the camera, and others would stand far away. Based on these photos, she created sketches around multiple bodies, which informed how using fasteners like drawstrings and magnets can connect people together in garments. She intentionally made some pieces that needed interaction and help from someone else. Her participants would ask, "The buttons are on

Figure 1.24 Models in the Beloved collection dressing together. Courtesy of Tiffany Hwang.

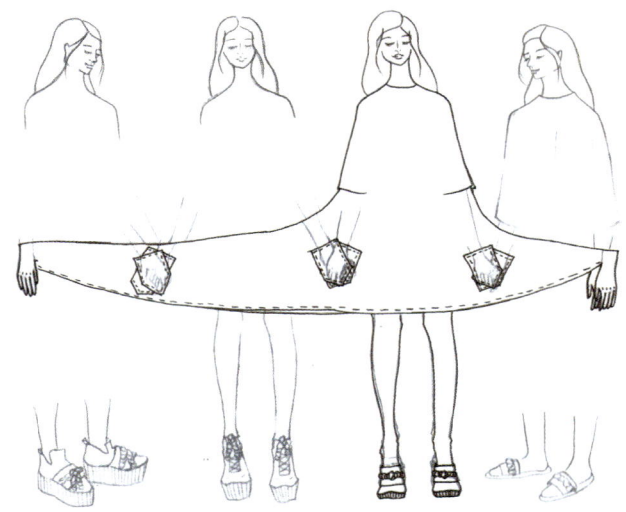

Figure 1.25a–b Sketches of designs and the dressing process by Tiffany Hwang. Courtesy of Tiffany Hwang.

Introduction 19

the back—can you reach them for me?" and "Can you come closer to fasten that (closure) on my garment?" These interactions reflected in Beloved are applicable to co-dressing and assistive dressing, demonstrating how fashion can transform the ways people get dressed into a collaborative experience.

Another design that could benefit more PWD is a series of bags by fashion and communication designer Jin Ah Jun, who demonstrates how a bag can be accessed from multiple openings using construction techniques and magnetic closures. The patterns in figures 1.26 a–b depict a side opening for two different bag designs.

Dressing Aids and Assistive Tools

Interdependent dressing involves other people but may also involve using assistive tools. With the impact of the Second World War and the polio outbreak, a rise in **occupational therapists**, often referred to as OTs, contributed to the craft, making, and education of assistive dressing devices or clothing solutions in tangent with the adaptive designs already explored by PWD. For example, hospital pamphlets and instruction manuals written by occupational therapists and medical specialists demonstrate early clothing modifications and dressing hacks. Early evidence from UK occupational therapist June Brown offers various adjustments that could be made to garments and articles of clothing to ensure that a

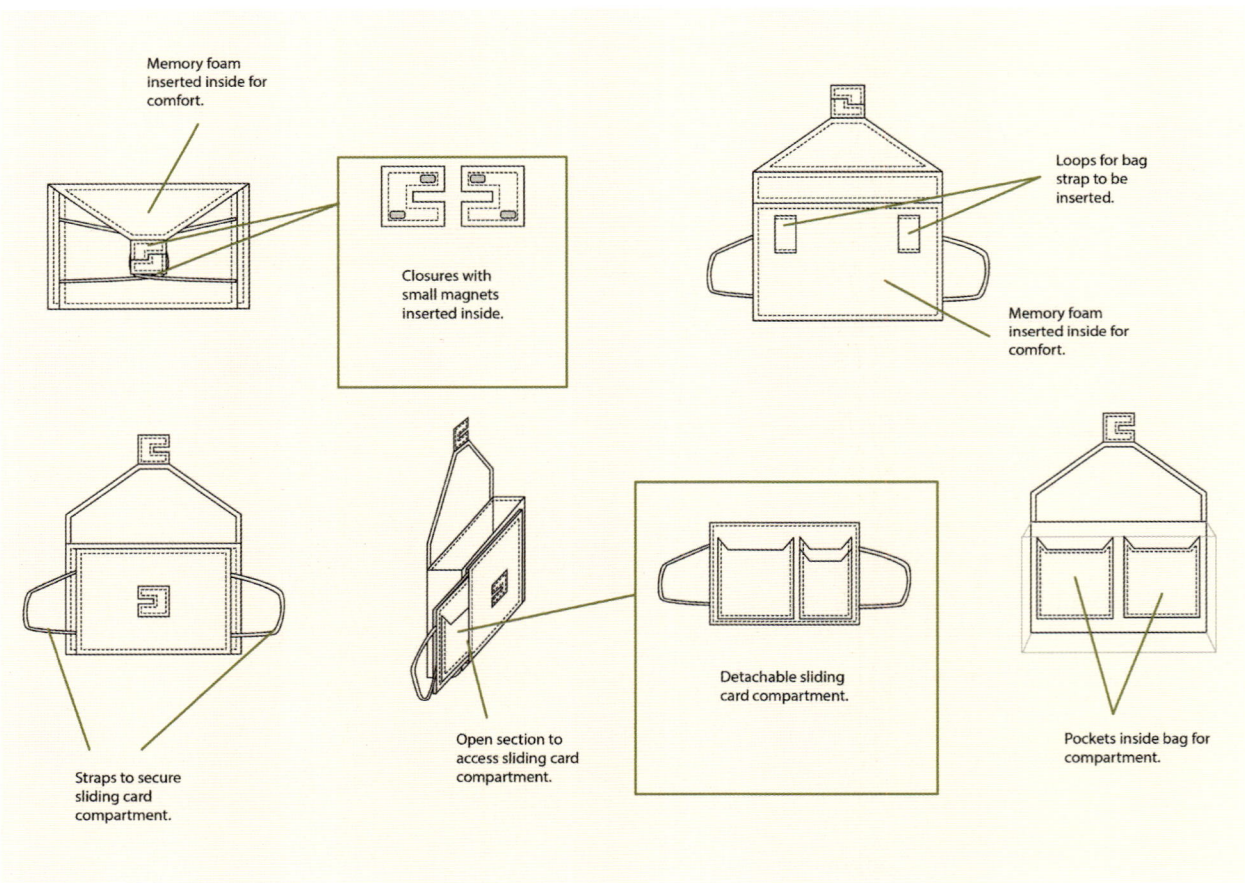

Figure 1.26 Sketches of two adaptive bag designs by Jin Ah Jun. Courtesy of Jin Ah Jun.

Figure 1.27a–b Christina demonstrating how she gets dressed. Photo taken by Unparalleled project collaborators, Julia Liao, Claudia Poh, and Estee Bruno. Courtesy of Julia Liao, Claudia Poh, and Estee Bruno.

person can become independent of other people while dressing (Brown 1959). Occupational therapists may also educate clients on adaptive equipment or tools for greater independence in dressing, such as sock aids (The American Occupational Therapy Association 2020). A sock aid features a rigid plastic or fabric shape that a wearer would pull the socks over and place it toward the floor to position one's foot inside while pulling up on the strings so that the sock will slide over one's foot (James 2014). Reachers, or devices that extend, are also popularized tools for assistive dressing used in occupational therapy and the disability community in various ways.

Tools used in occupational therapy correspond to ADLs (activities for daily living). This approach better informs fashioning designs for independent dressing and allows designers to prioritize a wearer's activity and environment. Identifying ADLs provides references for how people get dressed using tools and when maneuvering the body to adapt to situations. It encourages designers to contextualize where inaccessible dressing or fashion experiences may occur. For example, disabled designer Christina Mallon, who has little mobility in her arms due to paralysis from ALS, uses her bedpost as an assistive tool to hang her blouse while she thrusts her body and arms into the armholes to get dressed.

Christina uses her built environment to self-dress. From hanging her shirts on a bedpost to adding custom straps to her pants, Christina's hacks and methods provide a rich insight for non-disabled designers to reimagine dressing experiences, which is further explored in chapter 4, Unparalleled.

Conversation Points
- How will you define adaptive fashion?
- How will your design process include disability collaboration?
- How do design concepts generated by your co-design team provide benefits, feasibility, and novelty in fashion?
- In what ways can you identify accessibility needs in existing clothing?
- What dressing behaviors can you identify that will impact design choices?

References

Adam, Hajo, and Adam D. Galinsky. 2012. "Enclothed Cognition." *Journal of Experimental Social Psychology* 48 (4): 918–925. doi:10.1016/j.jesp.2012.02.008 (accessed August 14, 2023).

Allen, Danielle. 2016. "The Future of Democracy: How humanities education supports civic participation." *Humanities* 47 (2). https://www.neh.gov/humanities/2016/spring/feature/the-future-the-humanities-democracy (accessed August 14, 2023).

Berlinger, Max. 2022. "Adaptive Fashion Is in the Spotlight at Runway of Dreams in Los Angeles." *Vogue*, March 9. https://www.vogue.com/article/runway-of-dreams-adaptive-fashion-show (accessed August 14, 2023).

Brown, June. 1959. "A Guide Towards Helping the Disabled Person to Be Independent with Dressing." *British Journal of Occupational Therapy* 22 (3). doi:10.1177/030802265902200305 (accessed August 14, 2023).

Carroll, Katherine, and Kevin Gross. 2010. "An Examination of Clothing Issues and Physical Limitations in the Product Development Process." *Family and Consumer Sciences Research Journal* 39 (1): 2–17. doi:10.1111/j.1552-3934.2010.02041.x (accessed August 14, 2023).

Carroll, Katherine. 2015. "Fashion and Disability." In *Fashion Design for Living*, edited by Allison Gwilt, 151–167. Oxfordshire, England: Routledge.

Cook, Albert M., and Susan M. Hussey. 2002. *Assistive Technologies: Principles and Practice.* St. Louis: Mosby.

Cookman, Helen, and Muriel E. Zimmerman. 1961. *Functional Fashions for the Physically Handicapped.* New York: The Occupational Therapy Service Institute of Physical Medicine and Rehabilitation.

Dallas, Merry J., and Louise W. White. 1982. "Clothing Fasteners for Women with Arthritis." *American Journal of Occupational Therapy* 36 (8): 515–518. doi:10.5014/ajot.36.8.515 (accessed August 14, 2023).

Deihl, Nancy. 2018. *The Hidden History of American Fashion: Rediscovering 20th-Century Women Designers.* London: Bloomsbury Publishing.

Farra, Emily. 2017. "The Next Phase of Inclusive Fashion: Designing for the Disabled." *Vogue*, August 14. https://www.vogue.com/article/open-style-lab-fashion-showcase-parsons-designing-for-the-disabled (accessed August 14, 2023).

Feitelberg, Rosemary. 2021. "Parsons and the Special Olympics Working to Create Sustainable Change in Inclusive Apparel." *WWD*, May 7. https://wwd.com/fashion-news/fashion-features/parsons-special-olympics-nigel-barker-inclusive-design-adaptive-apparel-1234819093/ (accessed August 14, 2023).

Franklin, Ursula. 1990. *The Real World of Technology. CBC Massey Lectures.* Toronto: CBC Enterprises.

Freeman, Carla M., Susan B. Kaiser, and Stacy B. Wingate. 1985. "Perceptions of Functional Clothing by Persons with Physical Disabilities: A Social-Cognitive Framework." *Clothing and Textiles Research Journal* 4 (1): 46–52. doi:10.1177/0887302X8500400107 (accessed August 14, 2023).

Gupta, Deepti. 2011. "Functional Clothing—Definition and Classification." *Indian Journal of Fibre & Textile Research* 36 (4): 321–326.

James, Anne Birge. 2014. "Restoring the Role of the Independent Person." In *Occupational Therapy for Physical Dysfunction*, 7th ed., 753–803. Philadelphia: Lippincott Williams & Wilkins.

Keist, Carmen. 2017. "Stout Women Can Now Be Stylish." *Dress: The Journal of the Costume Society of America* 43 (2): 99–117. doi:10.1080/03612112.2017.1300474 (accessed August 14, 2023).

Kernaleguen, Anne. 1978. *Clothing Designs for the Handicapped.* Edmonton, Alberta, Canada: The University of Alberta Press.

Kerr, Marion. 1946. "Nursing Responsibilities in Cerebral Palsy." *The American Journal of Nursing* 46 (7): 469–474. doi:10.2307/3456670 (accessed August 14, 2023).

Lamb, Jane M. 2001. "Disability and the Social Importance of Appearance." *Clothing and Textiles Research Journal* 19 (3): 134–143. doi:10.1177/0887302X0101900304 (accessed August 14, 2023).

Lieber, Chavie. 2019. "The Adaptive Fashion Opportunity: Adaptive Clothing Is a $350 billion Opportunity the Fashion Industry Has Largely Ignored—Until Now." *Business of Fashion*,

October 22. https://www.businessoffashion.com/articles/technology/the-adaptive-fashion-opportunity/ (accessed August 14, 2023).

Loschek, Ingrid. 2009. *When Clothes Become Fashion. Design and Innovation Systems*. London: Bloomsbury Publishing.

Lupton, Ellen. 2014. *Beautiful Users: Designing for People*. New York: Princeton Architectural Press.

McBee-Black, Kerri, and Jung Ha-Brookshire. 2020. "Words Matter: A Content Analysis of the Definitions and Usage of the Terms for Apparel Marketed to People Living With Disabilities." *Clothing and Textiles Research Journal* 38 (3). doi:10.1177/0887302X19890416 (accessed August 14, 2023).

Morrow, Martha G. 1948. "Nylon Has Many New Uses." *The Science News-Letter* 53 (24): 378–380.

Quinn, M. Dolores, and Renée Weiss-Chase. 1990. *Simplicity's Design Without Limits*. Philadelphia: Drexel University Design Press.

Sagul, Edith A. 1943. "Procedures in Clothing Instruction in Classes for the Mentally Retarded." *Exceptional Children* 10 (1): 16–22. doi:10.1177/001440294301000104 (accessed August 14, 2023).

Simmel, Georg. 1957. "Fashion." *American Journal of Sociology* 62 (6): 541–558.

Spack, Harry. 1946. "Swimming Trunks US2524212A." *Google Patents*, patent issued October 3, 1950. https://patents.google.com/patent/US2524212#patentCitations (accessed August 14, 2023).

The American Occupational Therapy Association. 2020. "Occupational Therapy Practice Framework: Domain and Process Fourth Edition." *American Journal of Occupational Therapy* 74 (2): 1–87. doi:10.5014/ajot.2020.74S2001 (accessed August 14, 2023).

The Metropolitan Museum of Art. n.d. "Extreme Beauty: The Body Transformed." Exhibition Overview. https://www.metmuseum.org/exhibitions/listings/2001/extreme-beauty (accessed August 14, 2023)

Troy, Virgina Gardner. 2014. "Weaving Diplomacy: Textiles and Hand-Weaving at Home and Abroad at Midcentury." *Archives of American Art Journal* (53): 52–77.

Watkins, Susan M., and Lucy E. Dunne. 2015. *Functional Clothing Design: From Sportswear to Spacesuits*. London: Bloomsbury Publishing.

Williamson, Bess. 2019. *Accessible America: A History of Disability and Design*. New York: NYU Press.

Chapter 2
Theories, Models, and Methodologies

There are many ways to design, and this chapter serves as a guide on the adaptive fashion process. Frameworks, principles, methods, and techniques are showcased to provide designers with multiple ways of approaching adaptive fashion. Because adaptive fashion is interdisciplinary, the following theories and methods have been outlined to provide different perspectives. This chapter first explores existing fashions that have, or could have, the potential to be accessible solutions for disabled people.

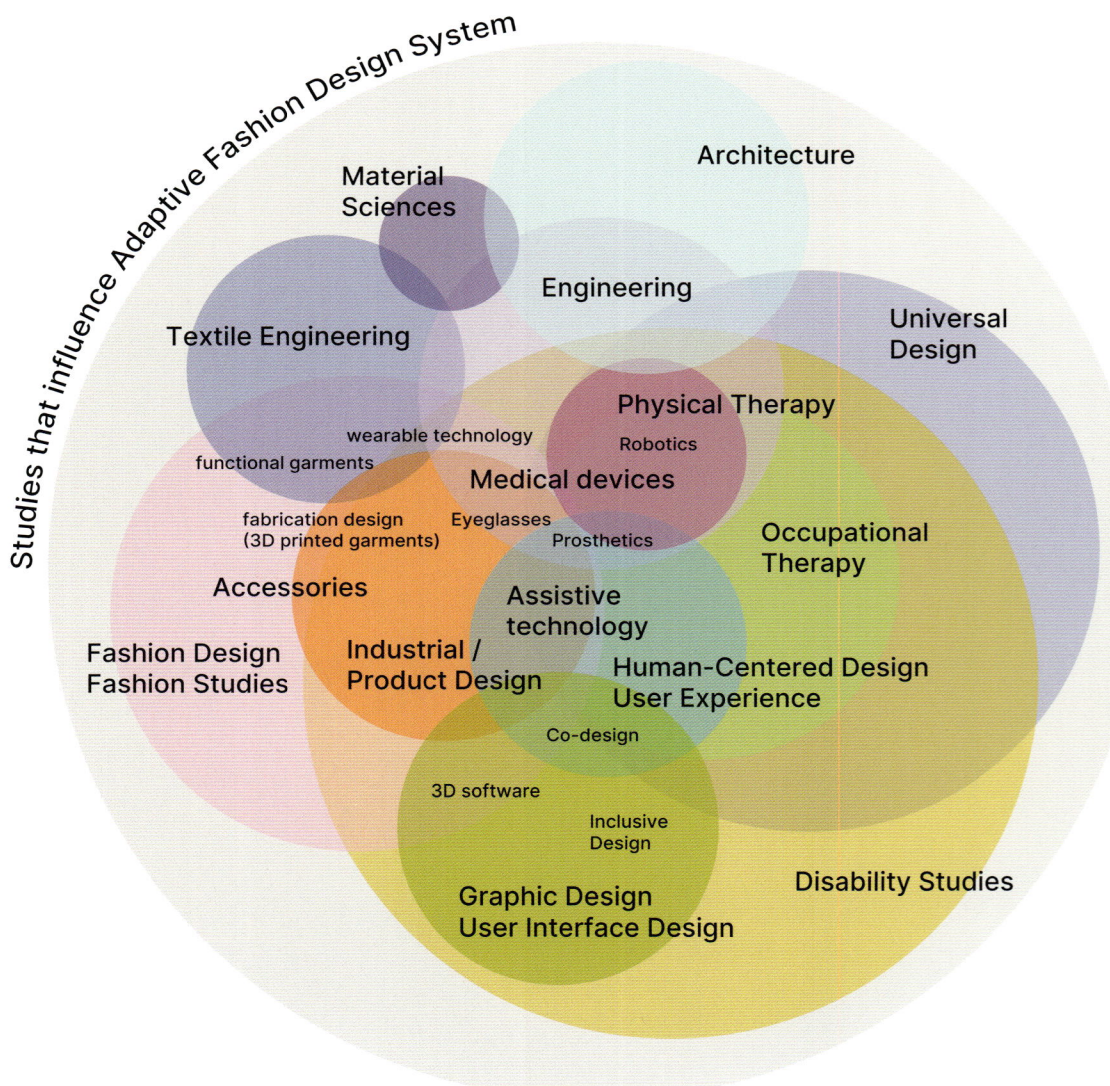

Figure 2.1 A diagram of intersecting studies that influence adaptive fashion design processes. Courtesy of Grace Jun.

Accessible Features in Existing Fashion Designs

Fashion designs that benefit disability needs have been presented in parallel with existing fashion houses, tailors, and practitioners at home. Designs that utilize functional materials have made it easier for more people to use and maintain garments. For example, wrinkle-resistant finishes provide easy maintenance and a prim, clean look and are particularly useful for PWD. Cleanliness and neatness were terms often used to describe a garment's desirability by people in the aging population, as further illustrated in chapter 4, LIULID.

Another fashion element that serves to provide a clean look with easy maintenance is pleating. Using a "garment pleating" technique, Issey Miyake launched a pleats line in 1988 that provided "light and wrinkle-proof clothing that doesn't need to be dry-cleaned and can be folded to a compact size for easy storage and carrying—with a versatility that makes them suitable for all settings in daily life" (Azzarello 2021). Length adjustable pleat dresses by Issey Miyake not only provide versatility to fit people of various heights and shapes but also choices in how the garments can be worn. Garments like these pleat designs by Issey are modular and transformative, exploring the use of shapes through intricate patternmaking and draping techniques, increasing their accessibility. This is also applicable to adaptive fashion designs that need to adjust to various body movements, postures, and shapes.

Utilitarian value isn't often associated with luxury fashions, yet there are notable designs that serve multiple functions applicable to disability. The Hermès spring/summer collection of 2000 features a belt pocket that provides adjustability and flexibility for the positioning of the hands—particularly useful for people who may have to rest their hands, such as the case of Christina in chapter 4, Swipe. An earlier example of Hermès

Figure 2.2 Issey Miyake, spring/summer 1994 by Issey Miyake, photographed by Niall McInerney in 1994. Copyright Bloomsbury Publishing.

in 1994 features fashionable front pockets on a burnt orange halter dress that is also applicable for wheelchair users, who may need to access pockets in the front.

These design examples from Hermès and Issey Miyake show luxury fashion's potential to be desirable and have a great functional use for disability. Transformative, reversible, packable, and modular garments, as seen in the 2019 Louis Vuitton collection by creative director Virgil Abloh, further show this. Abloh's recurring design details included bags that transformed into a garment. Another example of multifunctional clothing is Slick Chick, founded by Helya Mohammadian. The

Theories, Models, and Methodologies   27

Figure 2.3a–b Hermès, spring/summer 1994 and 2000, photographed by Niall McInerney. Copyright Bloomsbury Publishing.

brand's underwear designs can be put on when standing up or lying down, addressing the needs of people facing short-term or long-term disability and immobilization. In 2013, high-end fashion designer Hussein Chalayan created a printed cocktail-length dress that transformed into a long black dress, providing the option of different lengths for different occasions. Chalayan also challenges preconceived boundaries between body and space through the art of fashion performance. In figure 2.4, four models each take off seat covers that are then flipped inside out to be worn as clothing. In this example, the fashion experience includes the act of dressing while engaging with the built environment. This is often seen in adaptive fashion design studies, where disabled people are constantly engaging and adapting to spaces and materials surrounding them. For instance, dressing can be assisted by another person or even achieved by using furniture.

Rei Kawakubo takes this notion further by playing with proportion and disfiguring idealized body forms, as seen in figure 2.5 a-b. Design pieces like these deconstruct garments in a way that introduces new body types. Disabled people are not a monolith, and neither are their body types. Adaptive fashion design considers approaches like Kawakubo's work to unpack body image perceptions through clothing.

Figure 2.4a–b Hussein Chalayan, fall/winter 2000, photographed by Niall McInerney in London. Copyright Bloomsbury Publishing.

Figure 2.5a–b Comme des Garçons, spring/summer 1997 by Rei Kawakubo, photographed by Niall McInerney. Copyright Bloomsbury Publishing.

Theories, Models, and Methodologies

From couture to ready-to-wear, accidentally accessible designs have existed in many forms throughout fashion history yet are underutilized for disability needs due to issues in price or accessibility. The popularized wrap dress by Diane Von Furstenberg, also seen earlier in Helen Cookman's designs, is easy to put on and take off not only for able-bodied individuals but also for PWD (Cookman 1961). Yet, both designs have not reached enough consumers with disabilities. Part of that problem lies with not having enough disabled designers and collaborations involving disabled people. The following principles, models, and methods provide a guideline on how to leverage fashion practice to be more inclusive of disability.

Principles

Designers can utilize an array of design principles to examine the inaccessible barriers in clothing and accessories. Overall, the "essence of design is the process of discovering problems shared by many people" (Kenya 2018). Design principles such as UCD or human-centered design, universal design, **design for all**, and inclusive design are all various approaches to making fashion more inclusive. For the purposes of designing adaptive fashion for PWD, there are two key principles: inclusive design and fashion design.

Inclusive Design

Numerous scholars have investigated the human use of space and products, such as human-scale practices in architecture. These investigations question human normalcy and ability, which includes disability needs. One of the most influential principles that impact design processes addressing disability needs is inclusive design. Inclusive design is an approach that examines the diversity of experiences that may exclude a person from using a product, interface, or clothing. In applying inclusive design to fashion practice, one recognizes that wearable solutions for people with a disability are likely to also work well for people in diverse circumstances. For example, pants with pleats designed on the knee area targeted for seated wearers (i.e., wheelchair users) can also benefit people who do not use a wheelchair but are looking for a garment that is comfortable in both seated and unseated positions. Inclusive design champions including the end user (e.g., a person with a disability) in the process of making adaptive fashion. Therefore, an inclusive fashion design process must include research that explores scale: examining design approaches with one person with a disability, a specific focus group, and then with larger disability groups.

Inclusive design emerged in Europe during the late twentieth century as an approach to designing for all people. A similar approach called **universal design** championed by Ron Mace was developing in the United States. Both movements advocate design should be usable by all people to the greatest extent possible. Approaches like universal design have coincided with historic events such as the US Civil Rights movement in the 1960s, increasing awareness of the rights of all minority groups, which further integrates PWD into society. Along with the victories won by disabled activists, this influenced legislation like the ADA (Americans with Disabilities Act) in 1990 and amendment in 2008, which mandates that PWD have freedom, equality, and opportunity to participate fully in public life. Soon after, The ADA Standards for Accessible Design was published in 1991, which included a selection of laws and standards of Mace's seven universal design principles. While the practice of universal design has transformed design discourses around disability in the built environment (i.e., adding curb cuts to entrances or having elevators), little headway has been made in fashion or apparel. Disability in the built environment is seen as a spatial problem to solve. In fashion, the space between the human body and apparel is more intimate.

Inaccessible apparel can limit societal participation for PWD. Until recently, apparel design has been given little attention: The predominant focus has been on rethinking environmental barriers, such as inaccessible buildings or streets (Kabel 2016). Some design solutions using universal design principles were created with or by people who were historically excluded from participating in society, eventually becoming mainstream products and services. For instance, audio-recorded books created by blind communities or close captioning developed with people who are deaf or hard of hearing to make media more accessible.

Fashion Design

Fashion is more than just clothing. It is both a medium and message that is dependent on acceptance by observers and wearers alike (Loschek 2009). Fashion projects, therefore, creatively incorporate multiple aspects to articulate a story. It is a concept that is hard to grasp. The ambiguity of fashion expression sits uncomfortably between private territories of people's bodies and the public environment of perception. When developing an adaptive fashion project, the line-up of design components, such as technique or materials, is dependent on the scope: Is this project created to address an individual need or on an industrial scale for the masses? Fashion designers can convey a personalized story and capture nuances often overlooked in clothing, but scaling that message for a broader audience often requires a different design process.

Far too often, apparel designs inclusive of disability needs have been labeled as impractical in cost and serving a "niche" market (Casey 2022). Approaches to design for disability could be viewed as time-consuming, with made-to-measure needs. Yet other "niche" markets like plus-size wear have grown significantly in the United States, with a market size worth of $191.9 billion (Credence Research 2018). Furthermore, the practice of home-sewn fashions to create customized single garments has long been performed as an American tradition. With the rise of lighter and smaller weight sewing machines, home sewing became ubiquitous. Middle-class housewives and dressmakers practiced creating custom-made garments before the invention of paper patterns in the mid-nineteenth century (Kidwell 1974). A postwar shift toward conformity emphasized the importance for women to socialize and fit in as mass production of clothing was more readily available. The right kinds of clothing were critical to social acceptance in America, and plus-size and stout women in the 1910s were no exception. Despite stout clothing being available, "stout women were sadly neglected in the early development of the manufacture of dress" (Keist 2017). Fashion permeated throughout all aspects of society and was produced for individual needs and masses, yet disability was not fully recognized. Fashion studies and fashion marketing are, therefore, essential for designers to examine historic evidence and consumer behaviors that lack disability representation.

Finally, the vision and perspective of each designer are undeniably important to the adaptive fashion process. Fashion thinking, as described by Fiona Dieffenbacher, "is developing a creative and interdisciplinary mindset that goes beyond just making clothes" (Dieffenbacher 2021). Adaptive fashion designers are not exempt from needing to utilize questioning, visualizing, prototyping, and storytelling the process and outcomes found in any creative mindset. Like all fashion design practices, adaptive fashions must explore similar principles while integrating disability needs (FIT 2015):

- Balance—Balance is the visual weight in design and can be conveyed asymmetrically and/or symmetrically. All bodies are not perfectly symmetrical, yet many wearable designs do not include bodies that are drastically asymmetric. This principle offers designers to explore how a garment can provide a desired harmonious look regardless of body type and structure. An example of symmetrical design can be found in chapter 4, Zipback Jacket.
- Proportion—Proportion is the interrelationship between parts of an adaptive fashion design, including elements of scale, size, and fit. Size and fit will be the greatest challenge for adaptive fashion designers because PWD have varying proportions that do not conform to traditional body forms or standards used in schools and fashion shows. An example of size and proportion can be found in chapter 4, Modiste.
- Emphasis—A focal point in an adaptive garment can be achieved with color accents, contrast of shapes or details, and lines of fabric coming together.
- Rhythm—An adaptive fashion design with rhythm helps lead the eye from one part of a design to another part, creating movement through patterns using texture or color.

Additional design elements adaptive fashion designers should consider:

- Dynamic and Static Movement—People's bodies move differently depending on the context and environment. Designers can leverage construction techniques like patternmaking and draping to increase, restrict, and support body movements.
- Material Contact—Studying materials such as woven or knit textiles to discover design opportunities when in contact with the body is essential to any fashion study. When designing with PWD, materials play both a functional and aesthetic role. Some materials may cause or aggravate health conditions, while others may protect the body from weather ailments. An example that demonstrates material complexity is found in chapter 4, qXgo.

Processes, Models, and Methods

Processes

Co-design

Collaborative design, or co-design, can involve PWD early in the design process. Co-design research typically focuses on individuals or user communities, with the purpose "to empower participants to become legitimate and acknowledged members of the design team" (Visser 2005). Utilizing a co-design process increases a sense of ownership for PWD. PWD contribute to collective creativity for a fashion design process as experts in their lived experiences. Contrary to fashion design processes being the exclusive responsibility of design experts, co-design invites other people (e.g., end users, members, and customers) to become part of the team as "experts of their experiences" (Visser 2005).

The Slovenian-Croatian company, UCQC is an example of a design and disability collaboration, creating trendy yet affordable fashion for wheelchair users. Co-founders Hedvig Af Ekenstam and Maja Simunovic met at a Design (Dis)ability workshop in Ljubljana, Slovenia (see Warmed Bomber, chapter 4), where they created a jacket with a young woman, Katarina Milićević (Hackett 2021). Becoming tetraplegic after a car accident, Katarina wished to independently put on a jacket without any assistance from other people. To accomplish this, the team created several jacket prototypes, conducted market research, and worked with more disabled people, such as Paralympic athlete Luka Plavčak. One of the important features of the men's and women's jacket design was a row of loops along the zipper. A longer zipper at the hem of a jacket may be harder to reach with excess bulk or fabric. Loops that help with dressing easy-to-reach areas can be placed along the side of a jacket zipper, as seen in figure 2.8.

Co-design decentralizes a singular approach. Designing with disabled people results in highly functional design iterations that utilize advanced fabrication methods. Material and form are explored in tangent with body movement. For example, NuVu studio instructors Rosa Weinberg and Jenny Kinard collaborated with the renowned physically integrated dance company, Heidi Latsky Dance to co-design wearable art pieces with students and disability performers. Introducing students to disability artists and experts early in the design process, prior to any making, furthers collaboration.

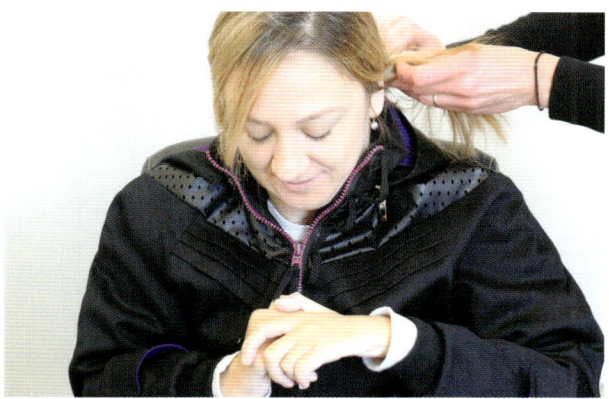

Figure 2.6 Katarina Milićević using zipper tag and loops when closing her jacket. Courtesy of UCQC Inclusive.

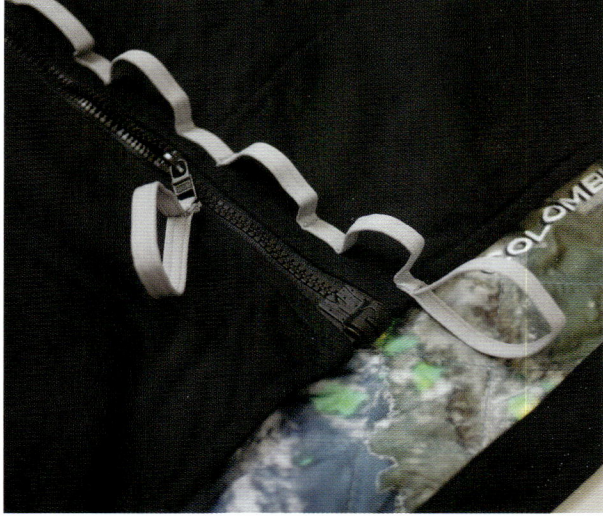

Figure 2.8 Men's adaptive jacket close-up with silver-colored loops alongside the zipper. Courtesy of UCQC Inclusive.

Figure 2.7 Men's adaptive jacket co-designed with Luka Plavčak. Courtesy of UCQC Inclusive.

Figure 2.9 Dancer Jerron Herman wearing the performative garment designed with Nina Cragg, Teresa Lourie, and Nya Rudek. Photographed by Amro Arida for Juxtapose (NuVu Studio) and Heidi Latsky Dance. Courtesy of Amro Arida.

Figure 2.10 Jerron Herman wearing the performative garment actuated by a pulling mechanism. Photographed by Amro Arida for Juxtapose (NuVu Studio) and Heidi Latsky Dance. Courtesy of Amro Arida.

One team of students created a wearable piece with Jerron Herman, a multi-talented professional dancer with cerebral palsy. The wearable piece is a shoulder strap with petal-shaped pieces that expand or contract with Jerron's body movement. The petal-shaped pieces were created by laser cutting a material called Rowlux®, a plastic-like material. Providing a way for Jerron to express making space for himself in inaccessible places, this wearable project was produced for a series of sculpture court-like performances called D.I.S.P.L.A.Y.E.D by Heidi Latsky, which ran at the Baruch Performing Arts Center in New York in 2018. Jerron describes the process as "enlivening—combining design and dance in such integral ways … Especially when you consider that these garments were handcrafted and intended for disabled bodies to perform, this process narrates about twenty years of progress in politics, theory, and art in one experience" (Herman, pers. comm.).

This co-design process involved multiple iterations before the final design, which included the following: researching origami shapes, creating small laser forms that expand and contract, and exploring paper applications to the body. Integrating various fabrication, making approaches, and perceptions about the human body helps convey the larger narrative that often clothing, dress, and wearable pieces communicate.

Co-designed garments made with PWD can be performative for not only artistic purposes but also scaled for a larger audience. For example, designers can discover fashion solutions unique to disabled and aging customers that are also desirable to all people. Fashion marketplaces or platforms such as Patti and Ricky (United States) and Adaptista (UK) demonstrate the potential for bringing in adaptive fashion brands created by PWD to a larger customer audience. As a founder with a disability, Marie O'Sullivan-Abeyratne created Adaptista in 2019 as an inclusive and web-accessible fashion retail platform. Adaptista features several brands that address various disability needs: "Many of the brands were designed with specific conditions and needs in mind, but could be used much more widely" (Webb 2022).

Figure 2.11a–d Design iterations created and photographed by Nina Cragg, Teresa Lourie, and Nya Rudek. Courtesy of Nina Cragg and Nya Rudek.

Theories, Models, and Methodologies   **35**

Model Name	Strength	Weakness
The FEA (Function, Environment, and Aesthetics) model developed by Lamb and Kallal in 1992 can be used to analyze wearer responses to specific garment types and general clothing needs.	Detailed in how fashion and design aesthetics may impact disability and specific garments. User-centric model that places human needs first.	Considers the environment in the context of fashion studies and not the overall activity the wearer performs. Excludes the importance of physical safety.
The MOHO (Model of Human Occupation) by Kielhofner and Burke in 1980 is a client-centered occupational therapy model examining how routines, roles, habits, and performance capacity (i.e., physical, cognitive, and perceptual abilities) of an individual and the physical or social environment enable occupational participation.	Rather than creating adaptations that might exist within a product, this model champions removing as many inaccessible barriers as possible. Dressing is also included in the model's exploration of an individual's social participation.	Designers may often need advice from occupational therapists and would be exposed to medical terminology. This model doesn't deeply integrate the aesthetics of functional garments as part of the dressing process.
The HAAT (Human Activity Assistive Technology) model by Cook and Hussey in 1995 centers on integrating the human, the activity, and the AT.	Like the MOHO model, the emphasis on activities allows designers to contextualize an adaptive garment. This model provides a connection between fashion and technology that is inclusive of disability needs.	This model is more applicable to occupational therapy. It is strongly focused on assistive technologies, which do not always include or relate to clothing design.
The Social Model of Disability was developed in the 1970s in the United States and further explored by Mike Oliver. The model proposes that disability results from the interaction between a person's characteristics and their unsuitable environment—not their medical condition.	This model provides a powerful perspective when integrating social and political contexts into fashion.	The model excludes design principles that are essential in making adaptive fashion.

Models

This section focuses on how conceptual models from different disciplines can be applied to creating fashions inclusive of disability. Rather than a single solution, each model has strengths and weaknesses for the design process and can be selected based on the needs of an individual project. The interdisciplinary aspect of co-design becomes evident when each stakeholder contributes to the process with a different perspective. The models, in the table opposite, offer different ways to approach adaptive fashion design through **interdisciplinary design**:

FEA (Function, Environment, and Aesthetics)
Context is everything. As Entwistle writes, dress should be a "situated bodily practice" (Entwistle 2000). When designing any fashion product, the environment and person must all be considered. Clothes are not lifeless shells but rather "hold traces of the people that made, lived in, and shaped them" (Carroll 2015). This model helps adaptive fashion design processes by applying contextual insight in relation to aesthetics and how clothing may function for wearers.

Fashion performs multiple functions. Functional clothing can be defined as wearable parts that are specifically engineered to deliver a performance or functionality for the wearer. From protective outerwear garments in hiking to firefighter uniforms, the design and development of functional clothing are driven by the choice of materials, the social, psychological, or physiological requirements of the user, and ergonomic considerations, such as sizing and fit (Gupta 2011). The FEA model is one method designers can use when creating fashions inclusive of disability needs by referencing three

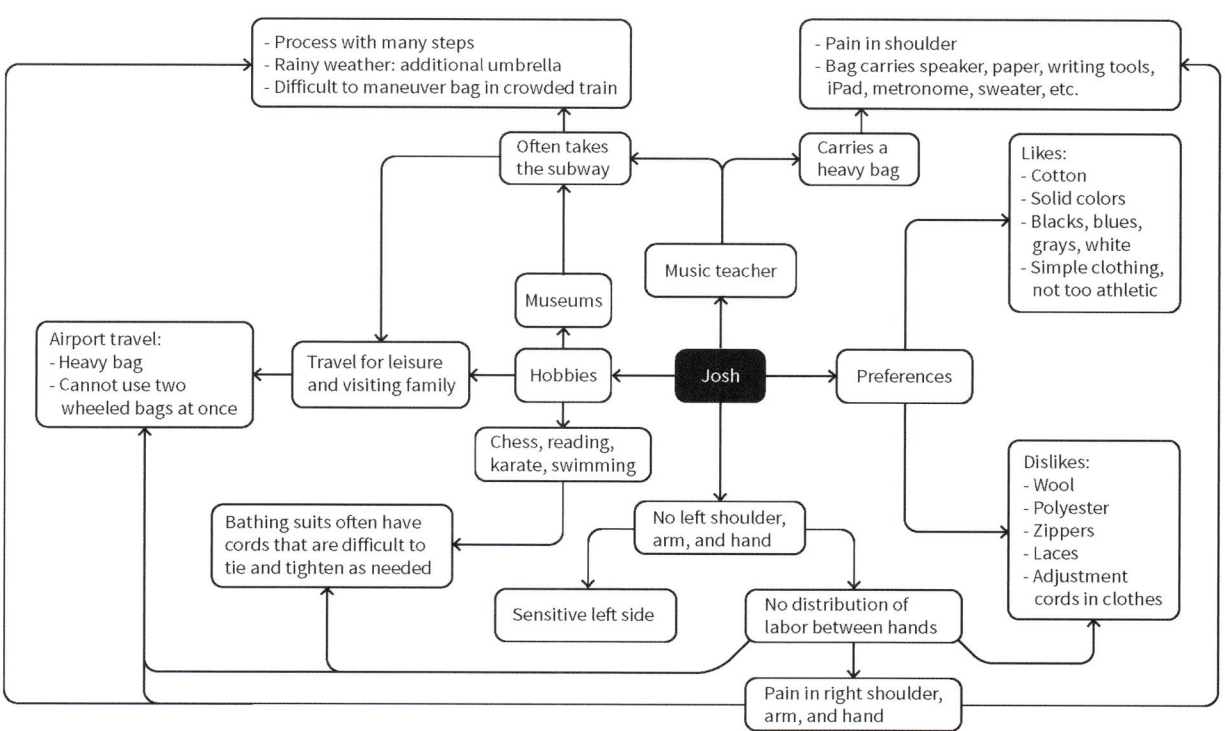

Figure 2.12 A journey map of Josh's city commute drawn by Célestine Yuan. Courtesy of Célestine Yuan.

Theories, Models, and Methodologies 37

 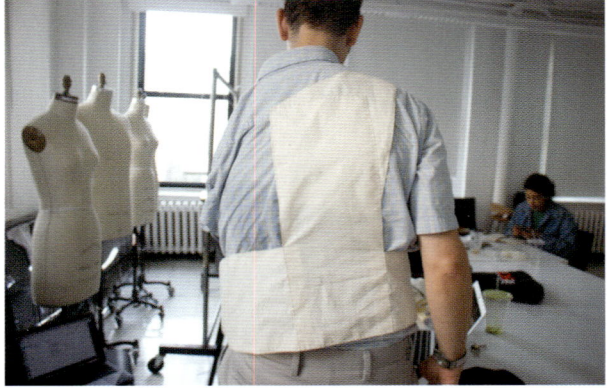

Figure 2.13a–b The front and back design of Joshpack created in muslin fabric. Courtesy of Open Style Lab™.

elements: function, aesthetic, and expressive (Lamb and Kallal 1992). Among the three elements include features of fit, mobility, comfort, protection, dressing on or taking off (donning and doffing), and aesthetic elements such as design principles that assist in observing the body or garment relationship. In this model, human behavior involving dress takes place within a social context. The model places the user (human) first. Body posture, social habits, and context are important considerations designers must keep in mind when creating types of clothing designs that would impact PWD. Observing how a person's body meets the environment and a variety of surfaces is essential to the adaptive fashion process. For example, designer Célestine Yuan sketched a journey route to better understand her collaborator, Josh Gilinsky, a music teacher and adult survivor of pediatric cancer with an amputation.

Célestine and Josh worked with two other designers, Zoey Zhou and Josefina Baillères, during an inclusive design class offered at Parsons School of Design. The team created the Joshpack, a sling pack designed for everyday single-handed use.

Figure 2.14 A body scanning mobile app used to gather data on fit measurements of Josh's arm. Courtesy of Camila Chiriboga.

Josh described finding a design opportunity to better distribute weight and provide access with one arm based on his shared subway ride with Célestine: "Carrying all these heavy music books to teach and to tutor, I sometimes have awkward size, paper, and scores. It's been a problem all my life. In transit, I couldn't turn my bag while having it still on to take out anything" (Gilinsky, pers. comm.). Single-handed use of clothing, accessories, and designs benefits not only Josh but all people. Mothers who need to hold a child while pushing a stroller could also benefit from the Joshpack. Since backpack measurements and shapes are crucial to how weight is distributed, the team used body scanning and 3D modeling to create an accurate pattern design for this customized bag.

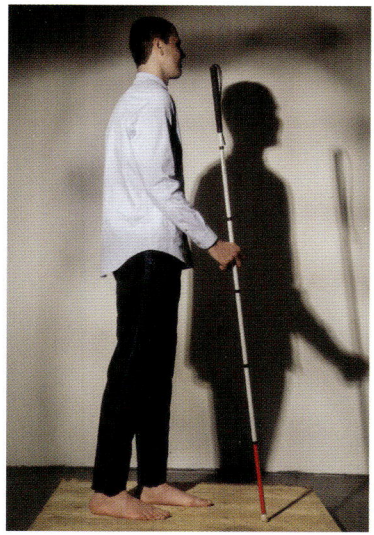

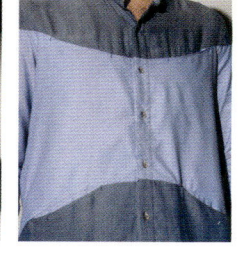

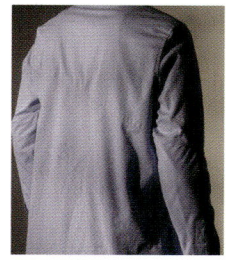

Figure 2.15a–e Sammi Hashash wearing a button-down shirt design from the Ve° collection. Courtesy of Camila Chiriboga.

Figure 2.16 Material textile exploration for garments created and photographed by Camila Chiriboga. Courtesy of Camila Chiriboga.

Technologies like this provide adaptive fashion designers with new tools and approaches to making. 3D printing has also provided an innovative method of production with customization, replicating shapes into textiles or forms. Prosthetics can be unique yet mass-produced at a scale with design choice. Tools like 3D printing or 3D modeling enrich adaptive fashion processes, as referenced in chapter 4, Swipe.

MOHO (Model of Human Occupation)

This client-centered model can help fashion designers research and incorporate the routines and habits of an individual. For example, designers can identify the habitual wear of a particular outfit. In theory, the model focuses on the entire human lifespan as the time frame for analysis. Objects that make up our environments are examined. Clothing or accessories not worn are arguably stand-alone objects. For example, a pair of shoes on display or belts hung in a decorative way and never worn. Designers can question how clothing can improve the quality of life for an individual, such as including long-term considerations like laundering and care. For example, designer Camila Chiriboga's project, Ve°, is a collection of clothes with special design elements and a tagging system that allows the visually impaired to choose their clothes. In the collection, button-down shirts were made from an easy-to-care-for material that did not require ironing.

The materials used had tactile textures that represented their visual aspect, for example, a striped shirt with lines that could be felt by passing a hand over it. All garments were reversible, preventing the anxiety or fear of wearing garments incorrectly.

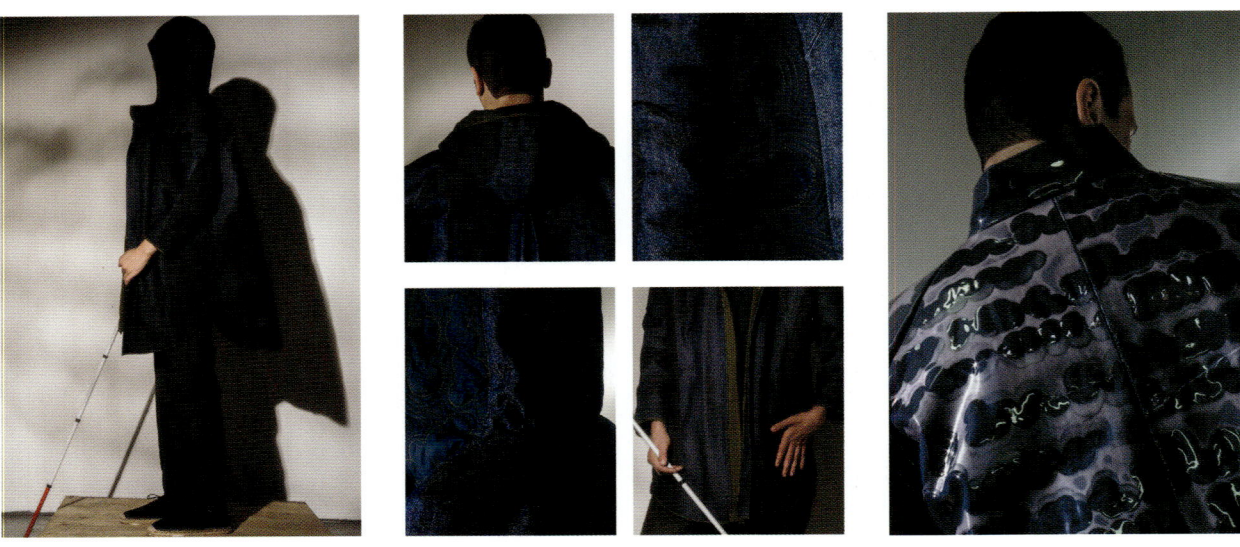

Figure 2.17a–f Heat Press Vinyl Jacket, a rain jacket heat pressed with an outline texture of braille, worn by Sammi Hashash and photographed by Andres Burgos. Courtesy of Andres Burgos.

Dressing is a different process for everyone. With reference to the MOHO model, Camila's design process included individual experience and close observation of the way the collaborators lived, worked, and the spaces they engaged in. Her design and process consider the longevity of a garment and the possibility of how it can impact the daily routine of the wearer.

HAAT (Human Activity Assistive Technology)
Practiced in the field of AT, in 1995, Cook and Hussey proposed a model framework for understanding the place of AT in the lives of PWD. Adaptive fashion designs that relate to AT can reference this model as a guide to determine what a wearer considers as part of the body, an extension of the body, or a product of the environment. This model is useful for adaptive fashion designers because it examines the interaction between device and person. It often helps designers examine clothing design choices as an interactive experience for the wearer, especially if it includes technology tools like electronic devices or sensors.

Disabled designer Larissa Sehringer created various 3D printed zipper-hook designs that could be more accessible for people who lack fine motor skills or experience limited dexterity in their hands. Her iterations for zipper design pulleys and hooks were informed by a specific activity: reaching a backpack when commuting in the subway. The purple backpack design used various stitching techniques that made a lightweight fabric durable. The 3D printed zipper-hooks offered an alternative way for wearers to open hard-to-reach pockets. This design demonstrates how design approaches in fashion products can have a close relationship with AT processes. Both studies look at examining the object and person in the context of a specific activity.

The paradigm shift of assistive devices to adaptive fashion includes products, such as canes and eyeglasses, that have long danced between

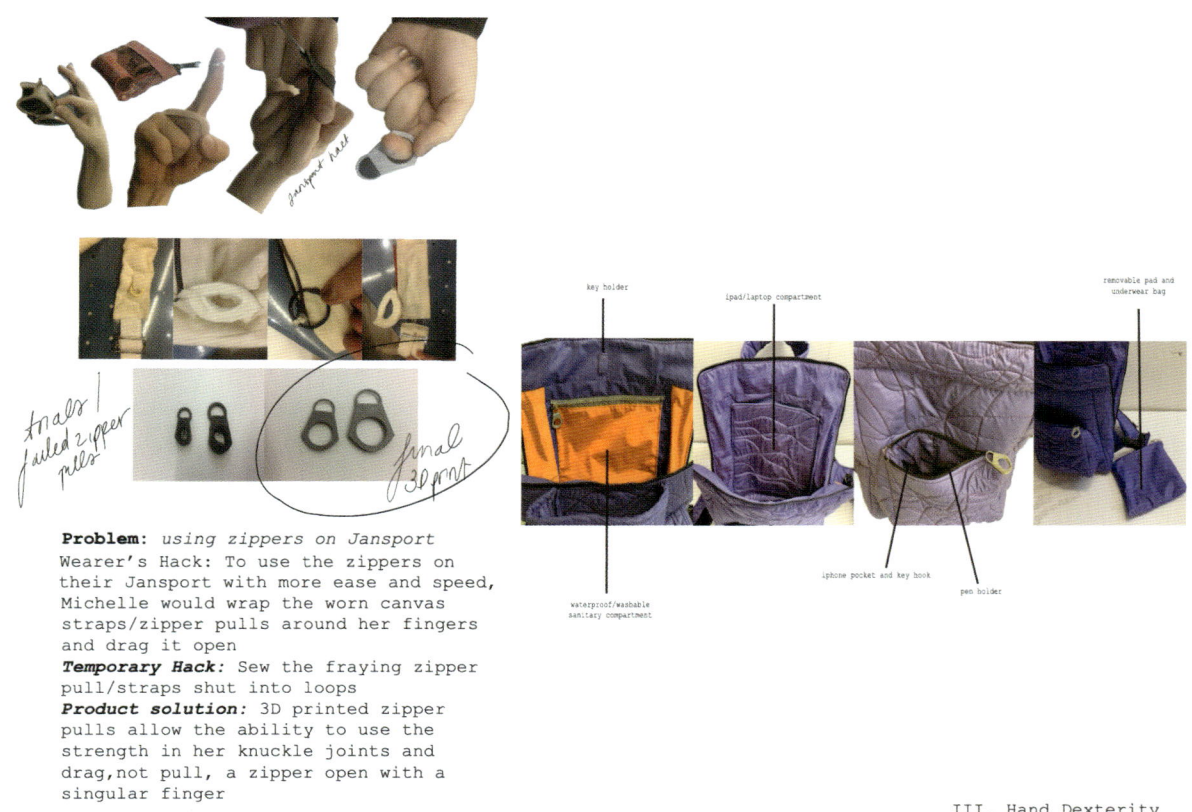

Figure 2.18 Iterative sketches and collages of zipper-hooks for an adaptive bag design created by Larissa Sehringer. Image courtesy of Larissa Sehringer.

the spectrum of being purely functional and a fashion statement. Glasses evolved into a fashion object for all due to "a conscious and concerted effort on the part of eyewear manufacturers," states Jessica Glasscock (2021). Her research reveals "trade associations of opticians, optometrists and ophthalmologists were advocating for control of the sales and manufacturing of corrective eyewear, effectively muscling the more fashionable department store and the jewelry counter to the fringes of the eyewear business" (Glasscock 2021). The evolution and success of eyewear from an assistive device to an accessible fashion accessory demonstrates the possibility of adaptive clothing.

Figure 2.19 Chanel, spring/summer 1996, photographed by Niall McInerney. Copyright Bloomsbury Publishing.

Theories, Models, and Methodologies   41

Researcher and writer Dr. Jo Gooding states how spectacles have become a "fashionable accessory that for many no longer has a stigma attached" (Gooding 2022). Designers can reference the relationships particular fashion items have held throughout a period that has transitioned between the medical and fashion realms. In doing so, designers can better identify the market and user scope of clothing items that involve specific practitioners, manufacturers, and distributors.

The Social Model of Disability
Disability studies greatly influences design approaches in adaptive fashion. Designs developed for and with disabled individuals eventually become conveniences for all people. Disabled people's own interventions in everyday designs have presented a parallel storyline for access alongside the history of clothing design. PWD have demonstrated creative ways of adapting, repairing, or inventing designs to be accessible. Some of the earliest inventions were created with and by PWD. For example, Alexander Graham Bell invented the telephone because his wife and mother were hard of hearing. Oxo grips were invented by Sam Farber, whose wife had arthritis. From telephones to kitchen tools, these designs demonstrate innovation that started out as inclusive solutions.

Yet historically in the United States, apparel designed for PWD featured rehabilitative attributes instead of fashionable attributes (Carroll 2015). A design that neglects aesthetics is often unsuitable for occasions with unspoken dress codes, such as wedding attire or workplace uniforms. Lack of appropriate clothing that integrates function and aesthetics, therefore, further excludes PWD. Aesthetics in fashion design can reduce disability stigma, allowing all people to live a socially active life. To accomplish this, adaptive fashion must meet a different way of thinking—a perspective that connects disability culture with fashion. At the core of disability studies, the field's **social model** translates insights of the United States and UK disability activities into an academic theory (Hamraie 2017). Popularized by Mike Oliver, the model suggests disability becomes apparent when there is an interaction, or commonly a mismatch of interactions, between a person and the objects, services, or spaces encountered. The social model of disability, when applied to the adaptive fashion design processes, supports the following: the diversity of people's experiences in dress and clothing as objects and interactive experiences; how bodies meet the designed world through fashion culture (The American Occupational Therapy Association 2020); making clothing accessible and inclusive of disability, prioritizing the idea of clothing as a cultural object and experience. The social model of disability implies how adaptive fashion functions beyond the physical but must also consider social implications such as occupation or relationships.

Method

UCD (User-Centered Design)
One way to evaluate integrating inclusion into the adaptive fashion processes is **UCD (user-centered design)**, a methodology that focuses on users' needs in each phase of the design process. Through an iterative design process, UCD uses various research methods that "facilitate user trials, considered to be a reliable way to identify usability problems" (e.g., Nielsen and Landauer 1993; Ebling and John 2000) within inclusive design in particular (e.g., Cardoso et al. 2005). UCD can focus on one or a few users for an in-depth design approach that can be highly personalized and customized to that user's needs and wants. User-centered design approaches can help differentiate between various disability impairments and further people's understanding of disability.

An example of UCD is practiced by Open Style Lab™, a nonprofit organization that aims to make style accessible for more people through educational programs and to raise awareness of the lack of accessible fashion for disability. Launched in 2014 as a public service project at MIT funded by a grant from MIT Age Lab and Eileen Fisher, Open Style Lab™ explores adaptive fashion inclusive of people with physical or cognitive disability. To accomplish this, the organization offers a summer program that teams therapists, designers, and engineers with a person with a disability to discuss inaccessible dressing and create adaptive clothing. Open Style Lab™ uses a co-design UCD approach to include a variety of multidisciplinary techniques and tools, where some project outcomes have also included wearable technology, as mentioned in chapter 4, Zipback Jacket. From integrating electronic sensors to using high-performing textiles, each design outcome demonstrates a range of innovations possible for adaptive fashion.

A prime design objective in UCD is that usefulness constitutes usability and utility (Clarkson 2003). Examining efficiency is a UCD feature that can be used from start to finish in the adaptive design process. The example of Abby's and Mickey's collaborative ice-skating design process mentioned in chapter 1 demonstrates how designers can identify usability needs. Abby recalls the design process began with how they "focused on the [skating] practices and Mickey's anatomical movements for figure skating moves that were particularly challenging" (Gaskin, pers. comm.). This informed functional features, such as adjustable straps for quicker dressing changes and a seamless crotch area designed in the pants to support specific figure skating movements, such as skating backward. The detachable crotch also allowed for easier transitions when using the restroom.

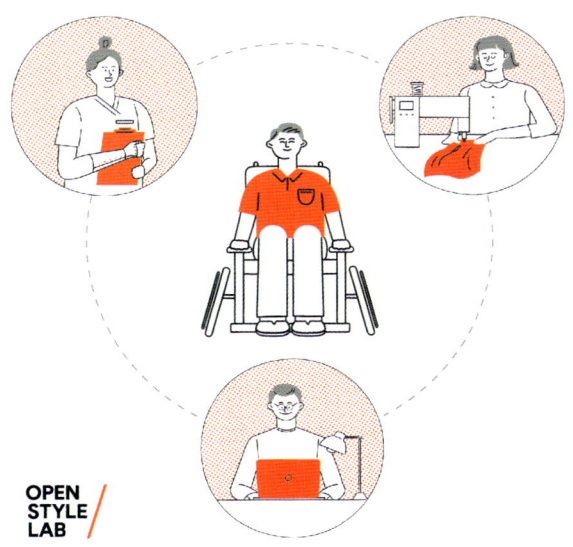

Figure 2.20 Open Style Lab™'s educational summer program model for interdisciplinary collaboration, where a person with a disability is teamed with an occupational therapist, engineer, and designer. Diagram illustrated by Rachel Gahyun Park, copyright by Open Style Lab™. Courtesy of Open Style Lab™.

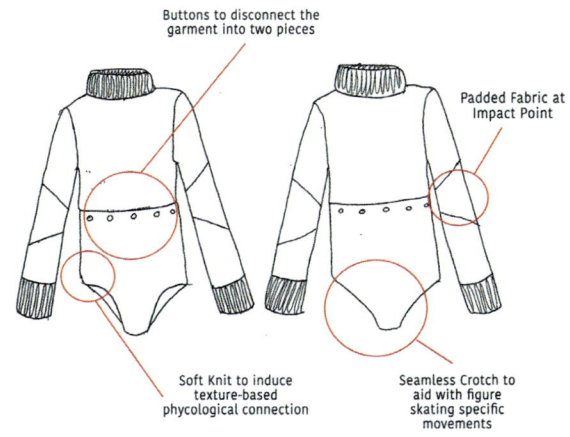

Figure 2.21 An adaptive jumpsuit with detachable features sketched by Abby Gaskin.
Courtesy of Abby Gaskin.

Theories, Models, and Methodologies 43

Human Factors

Consideration of human factors is essential in adaptive fashion design, influencing product design at different stages and levels. They are also particularly useful when assessing design performance. Since the study commissioned by the Human Factors Committee of the National Research Council in 1989, "human factors" has been defined as the term that covers both physiology and psychology, extending to elements that affect human performance. Using human factor methods in adaptive fashion design helps contextualize a type of activity and the inaccessible challenges of the built environment. For example, hearing, visual acuity, and temperature are some of the factors that directly affect human performance. Following this thinking, the ways each individual dresses or understands the clothing are also factors to consider in the design process. The perception of human factors is a mix of personal and public meanings that evolves over time and can be interpreted in different contexts. Adaptive designers who incorporate human factors in a fashion process could consider all physical elements of an apparel design that affect the person's comfort, safety, and performance, including the size and fit of a garment, tactile comfort, and visibility.

The development of adaptive design is a continuous interpretation and evaluation of specific human factors that are constantly moving from a subjective to an objective point of view. This occurs because the designer focuses on subjective and objective human experiences—"how the design of things influences people" (Sanders and McCormick 1992). While functional human factors in adaptive fashion can be interpreted clearly, aesthetic human factors are sometimes vague in their purpose because they embody multiple meanings depending on the wearer's use, intent, and style. Style is ultimately connected to each person, communicating individuality and personal character. Through choices in style, a person can visually show what cannot be verbally expressed. In adaptive fashion, the style of PWD can be explored using design methods and techniques. This allows an individual to communicate complex expressions of identity that include not only their disability but also age, gender, class, and religion. Identifying a person's sense of style and how that relates to the person's disability experience can be seen in chapter 4, The Little Black Bag. Finally, adaptive fashion design requires examining creative problems from a macroscopic lens before distilling theory into practice.

Conversation Points

- What methods will you include in your adaptive fashion process?
- What references in fashion design history better inform accessibility needs?
- What strategies can you map out to create your design process?
- How can you apply human-centered design or user-centered design processes?

References

Azzarello, Nina. 2021. "Unfolding the Art of Pleating: History + Techniques that have Fascinated the World of Fashion." designboom, 7 May. https://www.designboom.com/design/unfolding-art-pleating-history-techniques-fashion-05-09-2021/ (accessed October 26, 2023).

Cardoso, C., S. Keates, and P. J. Clarkson. 2005. *Are Users Necessary for Inclusive Design?* International Conference on Engineering Design (ICED), The Design Society, 15–18 August. https://www.designsociety.org/download-publication/23148/ARE+USERS+NECESSARY+ FOR+INCLUSIVE+ DESIGN%3F (accessed October 26, 2023).

Carroll, Katherine. 2015. "Fashion and Disability." In *Fashion Design for Living*, edited by Allison Gwilt, 151–167. Oxfordshire, England: Routledge.

Casey, Caroline. 2022. "Inclusive Design Is Not Niche." *Forbes*, October 3. https://www.forbes.com/sites/carolinecasey/2022/10/13/inclusive-design-is-not-niche/?sh=43b7d3041be7 (accessed August 14, 2023).

Clarkson, P. John, Roger Coleman, and Simeon Keates. 2003. *Inclusive Design: Design for the Whole Population*. New York: Springer.

Cookman, Helen, and Muriel E. Zimmerman. 1961. *Functional Fashions for the Physically Handicapped*. New York: The Occupational Therapy Service Institute of Physical Medicine and Rehabilitation.

Credence Research. 2018. "Plus Size Women's Clothing Market Size, Share and Forecast To 2026"

Dieffenbacher, Fiona. 2021. *Fashion Thinking*, 2nd ed. London: Bloomsbury Publishing.

Ebling, Maria R., and Bonnie E. John. 2000. "On the Contributions of Different Empirical Data in Usability Testing." The Conference on Designing Interactive Systems: Processes, Practices, Methods, and Techniques, Amsterdam: ACM Press. doi: 10.1145/347642.347766 (accessed October 26, 2023).

Entwistle, Joanne. 2000. *The Fashioned Body: Fashion, Dress and Modern Social Theory*. Cambridge: Polity Press.

Glasscock, Jessica. 2021. *Making a Spectacle: A Fashionable History of Glasses*. New York: Black Dog & Leventhal.

Gooding, Jo. 2022. "Re-framing My Vision for 2020." *Science Museum Blog*, July 5. https://blog.sciencemuseum.org.uk/re-framing-my-vision-for-2020/ (accessed August 2, 2022)

Gupta, Deepti. 2011. "Functional Clothing—Definition and Classification." *Indian Journal of Fibre & Textile Research* 36 (4): 321–326.

Hackett, Paul. 2021. "Functional Fashion for Wheelchair Users: The Social Enterprise Doing Good Business." *Euronews*, April 2. https://www.euronews.com/next/2021/04/02/good-business-the-power-of-social-entrepreneurship (accessed December 27, 2023).

Hamraie, Aimi. 2017. *Building Access: Universal Design and the Politics of Disability*. Minneapolis: University of Minnesota Press.

Kabel, Allison, Kerri McBee-Black, and Jessica Dimka. 2016. "Apparel-Related Participation Barriers: Ability, Adaptation and Engagement." *Disability and Rehabilitation* 38 (22): 2184–2192. doi:10.3109/09638288.2015.1123309 (accessed August 14, 2023).

Keist, Carmen. 2017. "Stout Women Can Now Be Stylish." *Dress: The Journal of the Costume Society of America* 43 (2): 99–117. doi:10.1080/03612112.2017.1300474 (accessed August 14, 2023).

Kenya, Hara. 2018. *Designing Design*, 4th ed. Zürich: Lars Müller Publishers.

Kidwell, Margaret C., and Claudia B. Christman. 1974. *Suiting Everyone: The Democratization of Clothing in America Paperback*, 1st ed. Washington, DC: Smithsonian Institution Press.

Lamb, Jane M., and M. Jo Kallal. 1992. "A Conceptual Framework for Apparel Design." *Clothing and Textiles Research Journal* 10 (2): 42–47. doi:10.1177/0887302x9201000207 (accessed August 14, 2023).

FIT (Fashion Institute of Technology). 2015. *Lauren Bacall: The Look*. 2015. "Elements and Principles of Fashion Design," edited by The Museum at FIT. New York: FIT.

Loschek, Ingrid. 2009. *When Clothes Become Fashion. Design and Innovation Systems*. London: Bloomsbury Publishing.

Nielsen, Jakob, and Thomas K. Landauer. 1993. "A Mathematical Model of the Finding of Usability Problems." Conference on Human Factors in Computing Systems (CHI '93), Amsterdam: ACM Press. doi: 10.1145/169059.169166 (accessed October 26, 2023).

Sanders, Mark S., and Ernest J. McCormick. 1992. *Human Factors and Engineering and Design*. New York: McGraw Hill.

The American Occupational Therapy Association. 2020. "Occupational Therapy Practice Framework: Domain and Process Fourth Edition." *American Journal of Occupational Therapy* 74 (2): 1–87. doi:10.5014/ajot.2020.74S2001 (accessed August 14, 2023).

Visser, Froukje Sleeswijk, Pieter Jan Stappers, Remko van der Lugt, and Elizabeth B. N. Sanders. 2005. "Contextmapping: Experiences from Practice." *International Journal of CoCreation in Design and the Arts* 1 (2): 119–149. https://studiolab.io.tudelft.nl/manila/gems/sleeswijkvisser/Codesign2005sleeswijk.pdf (accessed August 14, 2023).

Webb, Bella. 2022. "The Farfetch of Adaptive Fashion? High-End Marketplace Adaptista's Big Ambitions." *Vogue*, July 13. https://www.voguebusiness.com/fashion/the-farfetch-of-adaptive-fashion-high-end-marketplace-adaptistas-big-ambitions (accessed December 23, 2022).

Chapter 3
Approaches and Techniques

Adaptive fashions that are designed in an inclusive, collaborative way increase product compliance—keeping a design rather than discarding it. "Clothing items, which are perfect in comfort and function, may be completely rejected by the user if they do not look right" (Gupta 2011). Design led by or collaborated with PWD at the start of a process is crucial. The following approaches, techniques, and methods support incorporating human factors into fashion practice.

While all these approaches aspire toward a more accessible world, designers must integrate both knowledge and skill. In addition, designers need to consider the uncertainty of the design process with variability in situational contexts and use cases, further mentioned in chapter 4. They also need to integrate various expertise and go through rounds of experimentation to produce wearable outcomes for adaptive fashion. This ambiguous space of design experimentation is influenced by a paramount perspective: Our bodies don't follow predictable temporalities; they change over our lifetime. This idea is further framed as "universal disability" (Hamraie 2017). Clothing, therefore, is not a static design experience. Dynamic and changeable, clothing operates as a facilitator to enable, constrain, and organize bodies in different ways. This may restrict a person from moving a desired body part because of the material and construction of clothing. For example, a smaller armhole in a dress shirt that is made of a fabric with no stretch could make it more difficult for someone experiencing paralysis in the arms to don it. Performance factors (e.g., dressing behavior) intertwine with aesthetics (e.g., fabric color) when approaching adaptive fashion design. This chapter discusses guidelines, methods, and techniques for inclusive fashion design to serve as a blueprint while referencing the principles, models, and evaluation methods earlier in chapter 2. Being inclusive of disability in the design process is complex, intersectional, and interdisciplinary because it asks designers to integrate multiple perspectives beyond their own. Designers should consider the following approaches before starting a project:

Adaptive Fashion Framework

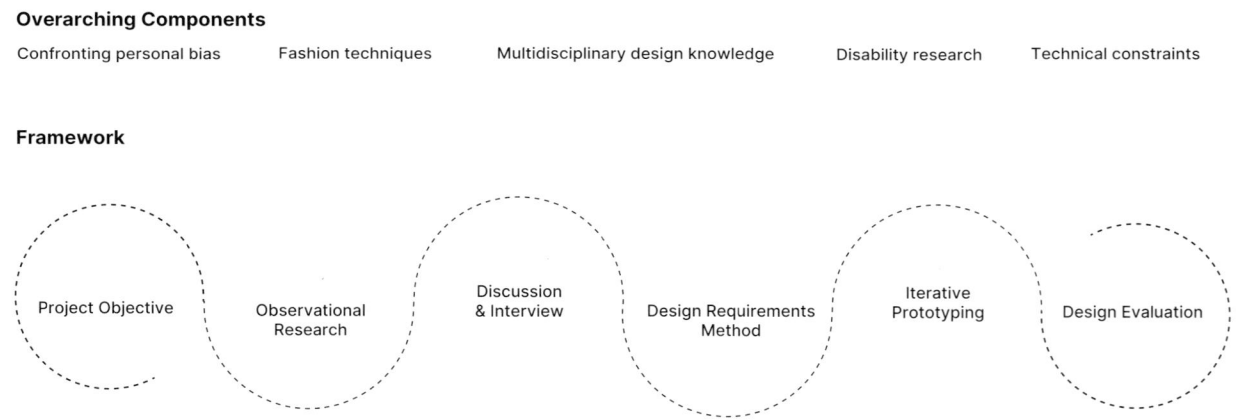

Figure 3.1 Adaptive Fashion Framework. Courtesy of Grace Jun.

Overview

- **Confronting personal bias**: While inclusive, universal, and co-design are all aspiring aims, identifying why and how people cannot use wearables (i.e., clothing, accessories) is critical. Products that are wearable, such as adaptive fashion, already present challenges that include materiality, body movement, and subjective opinions. For example, personal meanings are often assigned to clothes that have importance. The motives people have to use, wear, and interact with certain clothes help designers understand usability needs. Designers can then use suitable metrics and collect evidence that is quantifiable to identify how motives connect to design choices. For example, frequent use of a clasp may wear and tear over time, so a designer may research durable materials.
- **Fashion techniques**: Experience-based learning and making practices are essential to making any fashion piece. Adaptive fashion is no exception. An understanding of seam allowance, detailed hemming, and complex pleating are all examples of the variety of techniques a designer can utilize to make clothing more accessible for disability needs. A combination of construction techniques, materials, fabrics, patternmaking, and draping should be explored.
- **Multidisciplinary design knowledge**: Having the skills to make clothing is just as important as the ability to utilize different design approaches. Approaching designing for and with a disability should integrate methods of co-design, participatory, service design, universal design, and/or inclusive design.
- **Disability research**: Designers can refer to several books written by and with PWD on accessibility and inclusion (see Further Reading). This not only includes design methods and historic readings, but active field research to actively engage with disability groups. Designers should explore the medical, social, and identity paradigms of disability prior to, or during, the making process.
- **Technical constraints**: Understanding design constraints can allow designers to expand from an individual case of adaptive fashion to a broader group of people with and without disabilities. Creating an information architecture or design framework sets the boundaries in which the design will be constrained. This often includes the user's activity such as the environment, usability, and context for the overall design. An example of a framework designers can use is seen in the section called Design Requirements Method.

Guideline

Because designing adaptive fashion is multidisciplinary, some methods may happen in parallel with each other. While there are many ways to approach a design problem, below are suggested steps that can intertwine, intersect, or repeat throughout the adaptive fashion design process.

i. **Project Objectives**
   › Establishing a common form of communication
   › Making an objective

ii. **Observational Research**
   › Observational shadowing
   › Identifying repeated actions, habits of use, and/or dressing habits
   › Storyboards or visualizations that convey personal style
   › Research on ableism, individual style, and personal experiences on disability
   › Understanding lifestyle and unmet disability needs

iii. **Discussion and Interview**
   › Focus group discussions and/or diary studies
   › Qualitative or quantitative methods for interviews
   › User scenarios and user journey map
   › Analysis of existing clothing challenges
   › Contextualization of findings with larger groups (i.e., women, sports) by creating a visual map

iv. **Design Requirements Method**
   › Prioritizing a hierarchy of design choices
   › Functions to aesthetic features mapping
   › Examining how the adaptive design interacts with the user's body
   › Analysis of the design's potential to scale and apply to other use cases for broader user groups
   › Defining the problem

v. **Iterative Prototyping**
   › Rapid making and documentation using video, photography, and other visualization methods
   › Relationship and scale diagrams
   › Material exploration—the material properties of fabrics can be extremely complex and difficult to predict
   › Identifying technical and design constraints in fit and proportion
   › Identifying a user's range of motion and other movements that impact the dressing experience

vi. **Design Evaluation**
   › Qualitative and/or quantitative measures for design performance
   › Material selection and testing
   › Feedback assessment and reflection on initial objectives
   › Refining documentation material of the design process
   › Assessment of project limitations and possibilities

Application of Framework

This section explores how some of the processes above apply to adaptive fashion using an example project designed with Dorothy Jones called Access & Closure. Dorothy is a breast cancer survivor living in New York and, like many women, desires stylish blazers. Having breast cancer, Dorothy faces difficulty when wearing tight, form-fitting jackets. Dorothy, as well as other women with breast cancer during the earlier stages of recovery, have limited mobility in the arms and torso, making it difficult to put on a jacket.

This project focused on creating seven fashionable jackets with Dorothy. Each jacket in the series presents functional and aesthetic solutions for different phases of her recovery process. One of the jackets includes technologies that allow it to become an interactive communication tool to better inform her **physical therapist** of her daily symptoms. Another design feature includes the use of **conductive material** in the pockets that can be activated if the wearer has enough range of motion in the arms to press the pocket.

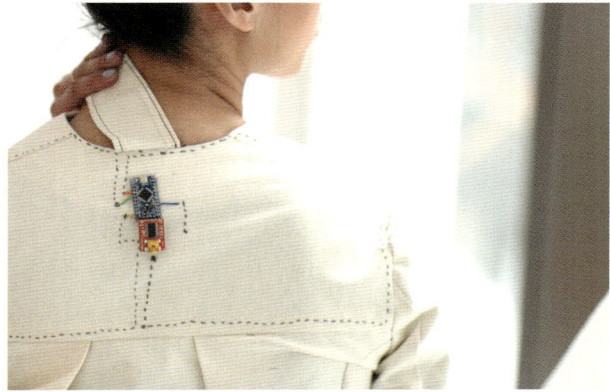

Figure 3.3 Muslin jacket prototype with circuit design to detect the range of Dorothy's arm motion. Modeled by Karen Chan and designed by Grace Jun.
Courtesy of Grace Jun.

Figure 3.2 Dorothy Jones wearing one of the seven jacket designs by Grace Jun, featuring a wide arm hold, elastic hooks on the cuffs, and an inside pocket for a hot or cold pack to relieve recurring chest pain.
Courtesy of Dorothy Jones and Alex Tosti.

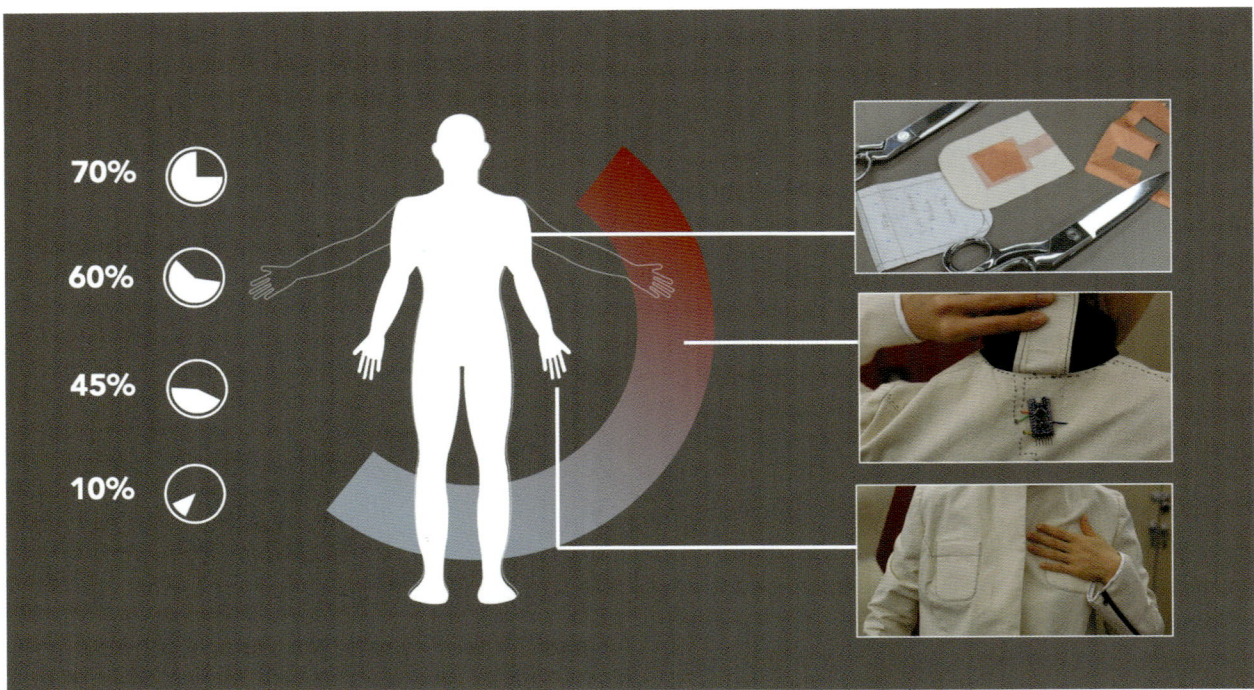

Figure 3.4 Range of motion diagram with circuit design embedded in the back of the jacket. Jacket designed by Grace Jun for Access & Closure collection. Courtesy of Grace Jun.

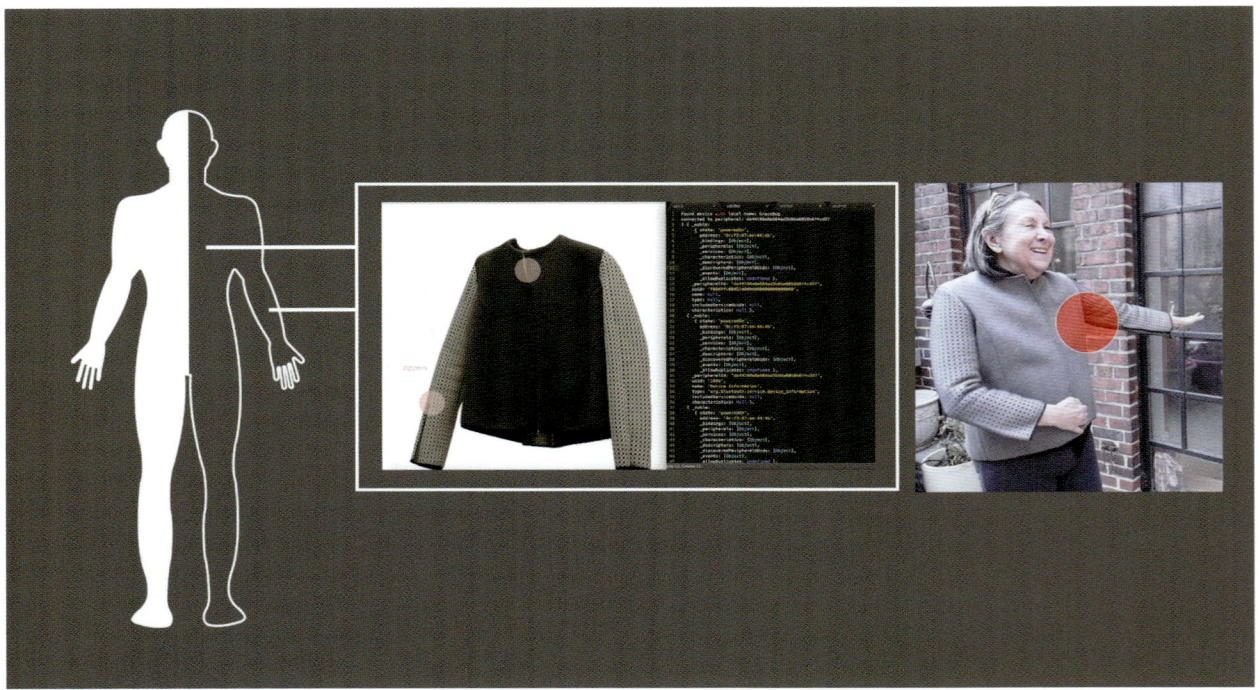

Figure 3.5 Conductive material spots for Dorothy to reach when demonstrating to her physical therapist her range of motion. Screenshot of code and jacket design developed by Grace Jun. Courtesy of Grace Jun.

To improve the wearing experience, these jackets were designed to include appropriate closures, fit, fabrics, materials, and style. For example, each jacket used wool blends, neoprene, and elastic polyester blends. The designs included a pleated back to support an increase in arm mobility, zippers and hooks on the sleeves for adjustable openings, and a wider armhole for easier dressing.

Project Objectives

Establishing a Common Form of Communication
With any co-design process, empathy is crucial. Addressing **ableism** is necessary for an inclusive design process. Establishing mutually agreed ways of communicating disability, body, and design vocabulary is therefore essential at the start of any design initiative. While Dorothy identified as a woman and a breast cancer survivor, the ways in which PWD wish to describe themselves may include the following:

- Person with a disability
- Disabled person
- A human being/person
- Preferred name
- Preferred gender pronouns

For designers without disabilities, starting a conversation on disability is often met with uncertainty. As Emily Ladau writes in *Demystifying Disability*, disability goes beyond inability, and it's recommended to first "ask people to share their preferred terminology" (Ladau 2021). The key to successful interviewing is learning how to probe effectively. To administer a questionnaire without subtly telling the respondent the answer you expect or over-injecting oneself into the interaction. Each data-collection method—face-to-face or video interview—has its advantages and disadvantages. The disadvantages of face-to-face interviews are that they are timely, costly, and can be intrusive and reactive. Yet, in-person interviews offer more detail about intimate activities such as getting dressed. In adaptive fashion design, questions about dressing processes are important to focus on first. Below are suggested questions to ask when communicating about dress and disability. The example questions and observations are based on an observational study that was used in the project with breast cancer survivor Dorothy Jones.

> Question 1: How do you describe yourself?
> Question 2: Do you identify as having a disability? If so, how would you like to be addressed?
> Question 3: How do you get dressed?
> Question 4: At what point do you get dressed? (e.g., after eating).
> Question 5: What items do you wear most often and why?
> Question 6: What are some of the most beneficial adaptations, techniques, or assistive devices that you have utilized?
> Question 7: What would you like to wear but is not accessible?
>
> Observation 1: Is the dressing performed independently? Are assistive devices or other props being used? Is someone assisting the person in getting dressed?
> Observation 2: Is dressing or taking off clothing performed lying down, sitting, or standing?
> Observation 3: What body part first meets the clothing or accessory worn?
> Observation 4: What is the sequence or way the individual gets dressed?
> Observation 5: How long is the entire dressing process?
> Observation 6: Identify the color, material composition, or type of clothing and/or accessory chosen.

Figure 3.6 Grace Jun and Dorothy Jones examining zipper measurements for the jacket design. Photography by Alex Tosti. Courtesy of Alex Tosti.

Figure 3.7 Sketches and material exploration of blazer styles by Grace Jun. Courtesy of Grace Jun.

These questions should be used to collect data about content and process and should not be relied on to generalize for all disability needs. Asking a person with a disability to identify what information they used to judge the benefits and drawbacks of certain clothing can be documented using a form, as shown in the Recognizing Existing Clothing Challenges section. Some feedback can be tailored to personal tastes or habits. For example, the wearer may prefer a garment that has been worn for a particular look over other possible solutions simply because of attachment. Other feedback observations can include the wearer perceiving that a particular material has certain beneficial attributes even if it does not. It is important, therefore, to distinguish the perception of clothing versus the actual construction of silhouette, color, and material of the article of clothing itself. Finally, designers should measure the time it takes to don and doff the clothing design. The following list provides ways to outline such observations:

- Username and date of observation
- Description of clothing observed
- Notes and fast sketches of clothing design
- Material composition
- Dressing time (donning and doffing)

Making an Objective

An objective helps guide a design process. In adaptive fashion, an objective often supports a person to accomplish a task. For example, Dorothy wanted to be able to wear blazers easily and without them causing pain in her chest and arms when moving. In other cases, objectives could be purely about a desirable style that meets the functional needs of the person. Dorothy expressed blazers had a "professional yet refined style," as opposed to wearing ponchos (Jones, pers. comm.).

A particular style may help integrate or produce a sense of belonging for the wearer. Desired aesthetics provide incentives for people to own their items longer. Design creates a sense of ownership through desirability and avoids creating products that are only functional. The User Persona template shown in figure 3.8 helps designers list out the possible styles and disability preferences for a client.

User Persona

User Name _____   Date of Observation _____

Disability
_____
_____
_____
_____

Time Experiencing Disability
_____ (number of years)

◯ Permanent   ◯ Temporary

How does the user's disability affect body movement and dressing?
_____
_____
_____

◯ Mobility Assistance

If yes, what are they?
_____
_____
_____
_____

What type of style does the person describe having?
_____
_____
_____

◯ Personal Assistance (ex-caretaker)

If yes, who are they and what are the responsibilities?
_____
_____
_____

What type of style does the observer describe the user having?
_____
_____
_____
_____
_____
_____

Figure 3.8 User Persona template. Courtesy of Grace Jun.

Research

Observing Dressing Behaviors

Observation techniques, skills, and practice are critical for any designer to truly engage and listen to disability needs. Observations of morning routines and dressing behavior are often not experienced first-hand in collaborative groups. But experience is everything. Designers who can experience the actual process of a person with a disability getting dressed will have different approaches to design as opposed to obtaining information about the dressing process from another source. This field experience for students and designers is critical to identify pain points in dressing. Observing the sequence of dressing without guiding the client is important. The following is a checklist of dressing observations referenced when working with Dorothy:

Bed-Related Activity
- Moving off and on the bed (especially if transferring to a wheelchair)
- Sitting upright on the bed
- Rolling right to left/left to right

Dressing Activity
- Putting on or taking off shirts with buttons
- Putting on or taking off pants with zippers
- Removing braces, prosthetics, or other supportive wearable devices attached to the body when tying shoes or putting on pants

Home Activities (that may inform dressing)
- Turning the pages of a book
- Turning a doorknob
- Open and closing doors
- Using a switch (push or plug in)
- Holding a cup
- Using a cell phone

Creating a User Journey Map

A user journey map helps contextualize where and when dressing challenges impact a person. Location, situation, and time are useful factors when identifying scenarios in which adaptive fashion best supports the wearer or user. A template is referenced in chapter 2, Joshpack, and chapter 4, Trans-Skirt. Designers can use these examples to document user activities by identifying the time, occasion, and place.

Recognizing Existing Clothing Challenges

Identifying pain points in the dressing process allows designers to observe how clothing can be less restrictive and inaccessible. For example, Dorothy's daily routine consisted of walks outside, reading the newspaper, and occasional events involving dressing for fundraisers, events, or family gatherings. The most common items found in her wardrobe were jackets and blazers. The following example form contains an analysis of several jackets Dorothy owned.

Pain Points of Existing Jackets	Accessibility Opportunities
Elbow area is restricting or confining to arm movement.	Switch fabric to an elastic and soft material that provides increased space for a greater range of motion in arms.
Ribbed cuffs provided stretch but not enough grip when dressing with one dominant arm.	Add loops to the ribbed cuffs to provide extra support for the fingers when putting the jacket on.
Buttons were too small or had magnet fasteners that were too strong to pull apart.	Switch buttons to larger buttons, snaps that look like buttons, magnets, or eliminate the buttons altogether.
Shoulder and underarm areas were restricted in fabrics that provided structural support but had no stretch.	Change fabric to an elastic and soft material that provides increased space for a greater range of motion (Chromat 2021) and reduced shearing.
Back of the jacket was tight-fitted and often uncomfortable for movement.	Install a pleat or Velcro® on the back that exposes additional elastic material to increase arm rotation or movements in the back.
Sleeves in suit jackets were tight-fitted and chic but were too tight to place arms through without experiencing pain or discomfort.	Adding a possible closure at the end of the sleeves, like a zipper or pleat, would help don/doff sleeves easily.
Pockets were angled in a position that caused keys and other valuable things to fall out easily. Most jacket pockets were on the outside and not large enough for Dorothy's hands to easily access.	Remove the top flap of the pocket to be an open pocket for use and angle pockets 25 degrees higher from the waistline.
Curved neckline designs of jackets were confining and stiff.	Widen collar based on body measurements to reduce shearing near the neck.
Inside of garments were usually not lined with smooth materials other than leather jackets.	Line inside the jacket with soft or silky material to reduce friction on and sensitivity of skin.

Design Requirements Method

A critical component in designing inclusively for PWD is including them at the very start of the process. One of the ways to distill design information, such as observation, discussion, and iterative making, is using the Design Requirements method. Prior to this method, observations and discussions should be conducted with the person with a disability. This initial research can include photo documentation, quotes, and, most importantly, documentation about the person and their dressing needs. When using this method, designers can gather this initial research to distill into variables and constraints. Variables, as described below, are factors that can be changed, modified, or altered in any way. For example, the length of a shirt is subject to change depending on the amount of fabric needed and body measurements. Constraints, like in many design processes, offer ways to scope and frame design observations. For example, a constraint in adaptive fashion can be a wheelchair if the user must use the device all the time. The wheelchair is a design constraint that impacts the choices a designer must make.

Methodologies must have utility. This ensures that they are relevant, meaningful, and practical for the chosen design. Successful adaptive fashions are those that are useful—as indicated by breadth, depth, or regularity of use. Retentive experiences provide ongoing value and make people's lives easier. For example, daily wear of a shirt demonstrates ongoing value because of the high frequency of use. If the design is not repeatedly worn or used (specific use case), the occasion that the design is worn should help the

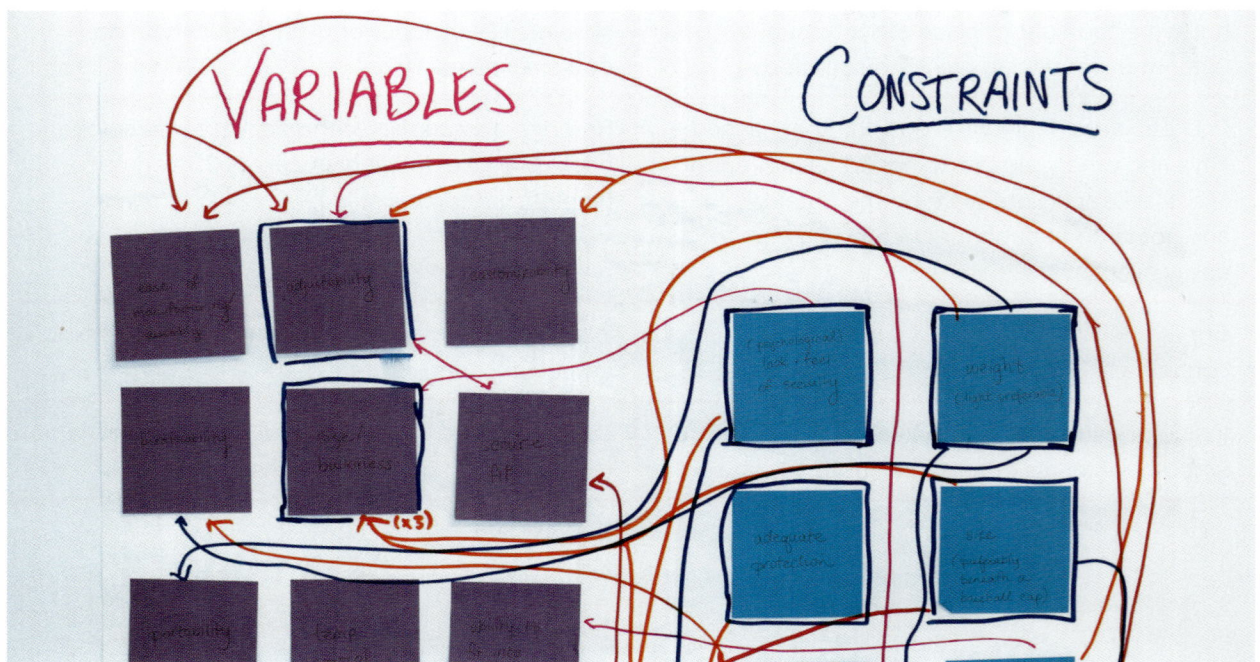

Figure 3.9 Idea sketch of variables and constraints in the design process. In this case the variables included: ease of manufacturing assembly, adjustability, customizability, breathability, size/bulkiness/secure fit, portability and temp control. While the constraints included: (psychological) look & feel of security, weight (light preferable), adequate protection and size (preferably beneath a baseball cap). Photo of sketch courtesy of Open Style Lab™, summer program 2015.

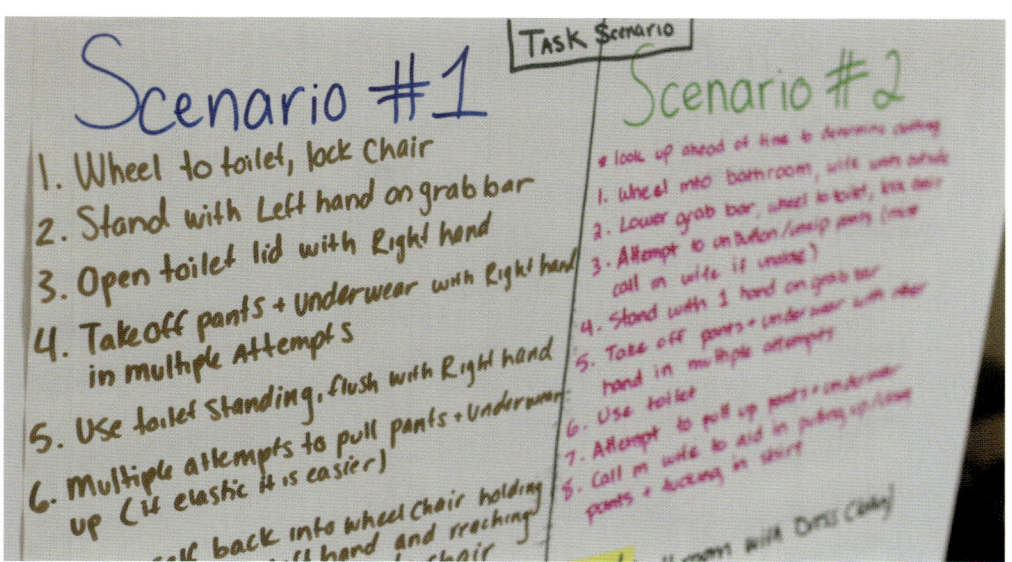

Figure 3.10 User Scenario sketch produced by Open Style Lab™ fellows in 2015 summer program. Photography courtesy of Open Style Lab™.

user overcome inaccessible barriers. For example, a suit jacket designed with and for a person using crutches to attend a wedding ceremony demonstrates a specific use case. The following steps offer a guide for designers to reference prior to the making process.

Design Requirement Method Steps:

- User scenarios
- Variables
- Constraints
- Framework
- Requirements
- Short summary statement

Design Requirement Method Goals:

- Defining a specific problem in inaccessible clothing and/or dressing.
- Organizing the research from the ethnographic observations.
- For people without a disability, being aware of biased opinions and personal experiences to avoid projecting biases onto PWD.
- Generating ideas from observations and connecting them to design choices.
- Developing a value proposition or unique selling point.

Step	Description	Example(s)
1	Gather initial observations of the person with a disability you are working with. Include significant quotes and documentation that reflect the person's style, preferences, and dressing processes.	Create a mood board that helps visualize the wearer's personal style with existing fashion references, including colors, trends, and materials. Take pictures of the dressing process if given permission. Write down quotes next to each stage of the dressing process.
2	Using sticky notes or other pieces of paper that are movable, make a list of variables and a list of constraints on each piece of paper.	Take ten sticky notes in one color and another set of ten in another color.
3	Create a list of words for variables. Variables are anything related to the user, context, or design that is changeable.	Adjustability Comfort Ease of donning and doffing Sleeve length  **Figure 3.11** Green sticky notes that list possible design variables. Courtesy of Open Style Lab™.
4	Repeat step 3 for constraints. Constraints are fixed elements. User constraints can be conditions that are not in the designer's power to change but may impact design choices.	Allergy to cotton Wheelchair Length of right leg Arm brace 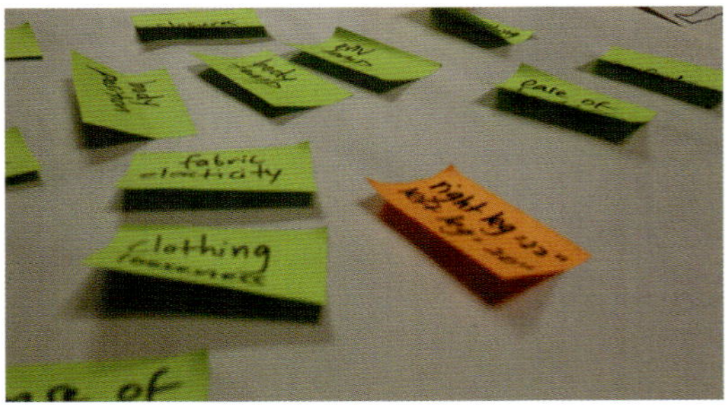 **Figure 3.12** Orange sticky notes that list possible design constraints. Courtesy of Open Style Lab™.

Step	Description	Example(s)
5	Prepare a large piece of paper or cardboard 24 x 18 in. (60.96 x 45.72 cm) to begin mapping out all the relationships between variables and constraints. This will help identify the most important design requirements.	 **Figure 3.13** Green and orange colored sticky notes placed on a large piece of paper to be organized. Courtesy of Open Style Lab™.
6	Gather only the pieces of paper that have variables. Using a marker, draw arrows to a variable that other variables are dependent on.	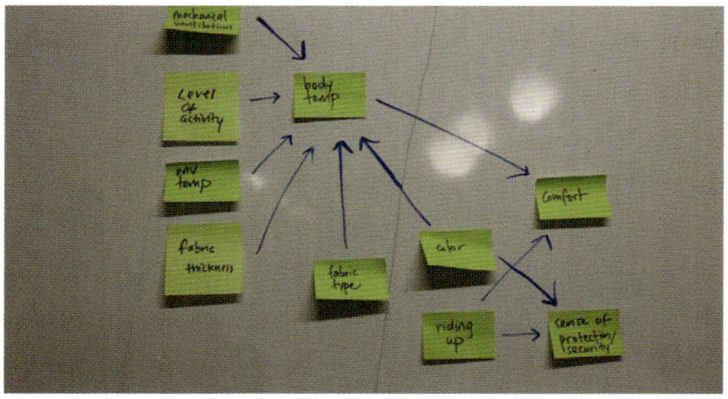 **Figure 3.14** A photo of only green sticky notes, or variables, listed with arrows on a whiteboard. Courtesy of Open Style Lab™.
7	Now add in the constraints and repeat Step 6. Note some variables may only point toward constraints, and some variables will point both ways.	 **Figure 3.15** A photo of Grace Jun placing all sticky notes and drawing arrows on a whiteboard. Courtesy of Open Style Lab™.

Approaches and Techniques 61

Step	Description	Example(s)
8	Find the elements (variable or constraint) that only have arrows pointing toward them and no arrows pointing away from them. Circle or mark those sticky notes with a different color.	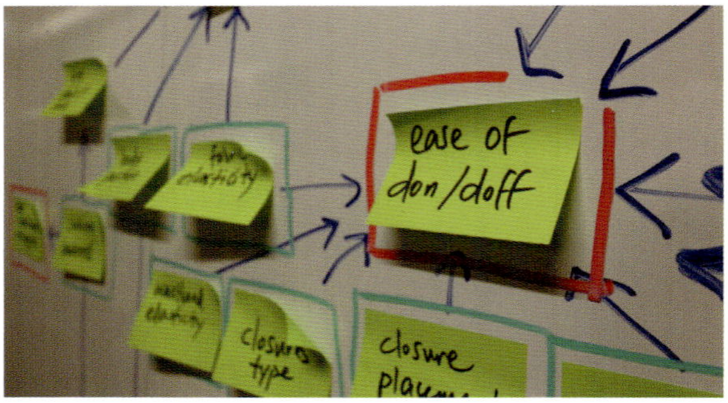 **Figure 3.16** Sticky notes with a marked red outline. Courtesy of Open Style Lab™.
9	List out the elements. These are the design requirements that the designer should refer to as success metrics during the making process.	Comfort Style Provides a greater level of independence in movement. 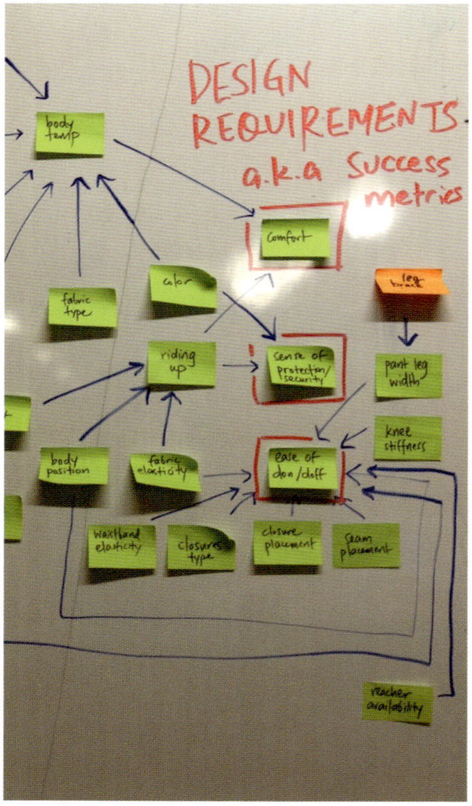 **Figure 3.17** Defining the design requirements. Courtesy of Open Style Lab™.

Step	Description	Example(s)
10	Circle the elements that have only arrows pointing away from them. Determine how to assess, test, or measure the element. These elements are factors that can help test theories and prototypes. This helps create an iterative design process to quickly test out ideas.	Fabric thickness Temperature of user's home Clothing looseness 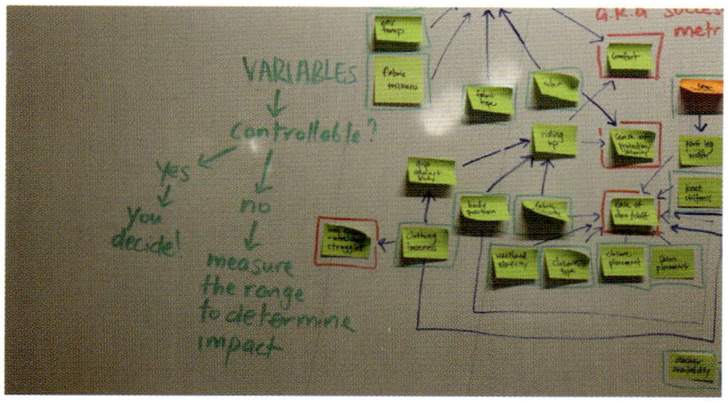 **Figure 3.18** Defining which variables can be tested. Courtesy of Open Style Lab™.
11	Create two to three user scenarios that narrate the dressing experience for the person in a particular environment or situation based on the quotes and connections derived from the list of variables and constraints. Drawings are also helpful in depicting each scenario.	1. User wheels into the restroom, with standby assistance from the spouse. 2. Transfers to the toilet seat while using the bar with the left hand. 3. Pulls shirt out of pants and unbuttons pants to pull down while seated for safety reasons. 4. Pulls elastic pants down. 5. To redress, the user pulls as much fabric as possible from the floor while stabilizing the left hand on the side of the toilet. Spouse comes to assist, and user transfers out of the restroom.

Note: All photos above were created and taken by Grace Jun and Dr. Grace Teo.

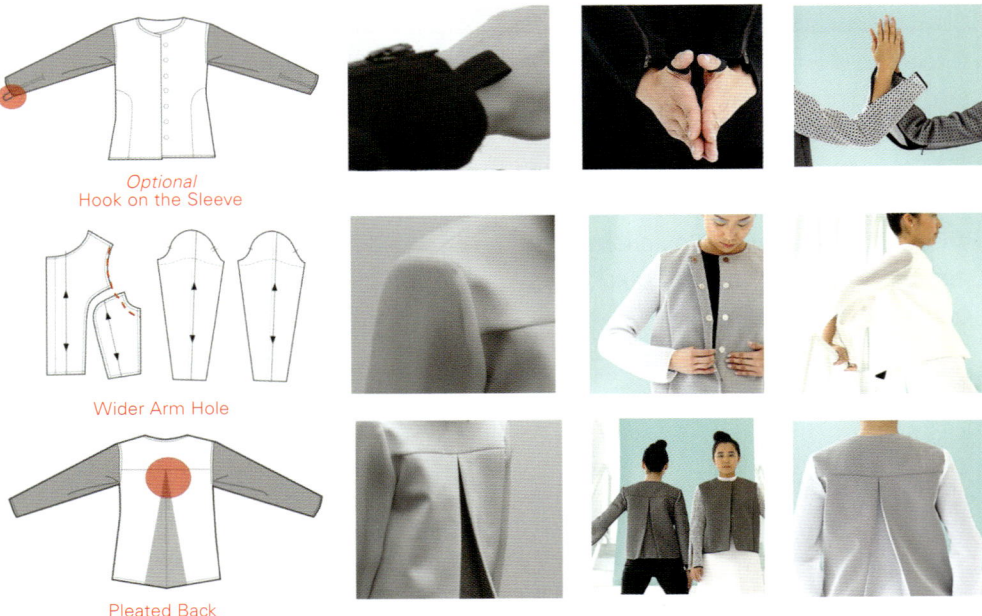

Figure 3.19 Pattern designs of jackets with wider armholes and pleated back, illustrated by Grace Jun on the left of the image. The design requirements are reflected in the construction and material of the jackets in the Access & Closure collection. Karen Chan and Jasmine Oh modeling collection, photographed by Alexi Tosti. Courtesy of Grace Jun.

Define the Problem

Using the Design Requirements method, the main pain points from the jacket project fell into the following categories:

1. Ease of dressing—lack of range of motion, variables that encourage independent dressing, and takes less time to dress. Garment designs that do not impede mobility or movement requirements and should be easy to put on or take off.
2. Protection—needs materials that provide absorbency when perspiring, breathability qualities, durability, or finishes that help regulate body temperature.
3. Comfort—challenges in fit, lack of material sensitivity, too much buildup of static electricity, and not inclusive of psychological sense of well-being for wearers.
4. Appearance—lack of choices in individual style and not related to fashion trends for social occasions and events.

Design Affordances

Design affordances refer to the possibility of an action or function of an object. They help determine the relationship between the person and the object, such as by examining the interaction of a silhouette or functional properties of a fabric. In the case of Dorothy, the following factors help evaluate, test, and measure the design factors that were created in the jackets:

1. Portability
2. Reach Support (extension or coverage)
3. Weight (light or heavy pressure)
4. Connectivity (conductive and communication)
5. Durability
6. Softness
7. Moisture/Water Wicking
8. Absorbent
9. Antimicrobial
10. Wrinkle-resistant
11. Moisture or Water-repellent
12. Ease of Donning and Doffing (dressing)
13. Movability
14. Comfort
15. Thermoregulation (cool and warm)
16. Discreetness
17. Attention Signaling
18. Breathability
19. Protection
20. Compression

Clothing Barrier	Affordance	Function	Feature	Application Example
Lack of designs with storage.	Portability	Provides multiple ways to store objects close to or on the body. Provides alternative ways for hands to be free.	Pockets.	Placing pockets in front of jeans instead of sides for people experiencing limited range of motion in the upper body.
Inadequate amount of fabric.	Coverage	Coverage for diverse body postures (e.g., hunched or seated).	Adjustable hems, extra fabric, pleats, overlapping panels.	Back of pants are too low for a seated user, so the back and the crotch area are extended.
Constricting styles for varied body movement.	Movement	Allows for greater range of motion for the wearer.	Actions pleats, underarm gusset, box polecats, yokes, back hem raised in jackets, elastic waistline, raglan sleeves, cut fabric on bias, bigger armholes.	Placing action pleats using stretchy fabric between the armhole and sleeves, allowing for greater motion when moving a manual wheelchair.

Clothing Barrier	Affordance	Function	Feature	Application Example
Challenges in dressing (donning and doffing) due to material or fabric of clothing.	Silhouettes for easier dressing	Allows for an easier dressing experience (donning and doffing). Prevents exertion and wasting energy or time.	Overhead, side, or inseam openings on bottoms or tops.	Using silhouettes or patterns that provide easier wearing, such as a boatneck style with a wider neck hole area. Raglan sleeve blouses also provide more room for easier donning and doffing a top.
Challenges in dressing (donning and doffing) caused by placement of clothing closures.	Clothing closures	Allows for an easier dressing experience (donning and doffing).	Easy-to-use fastenings such as magnetic zippers, longer loop tassels, loopholes, or bigger buttons.	Adding adjustable loops at the back neck area of a shirt for extra dressing support when the wearer faces difficulty in reaching far back to doff.
Preventing clothing from being revealing.	Weight distribution	Provides the wearer ways to adjust pressure with heavy weights.	Metal weights.	Placing metal weights at the hem of a skirt that are not magnetic, to keep skirt down when wind is blowing.
Lack of resistance to wear and tear.	Durability	Provides resistance to wear, tear, and overall protection of the body.	Durable fabrics and construction techniques that are abrasion resistive.	Designing an elbow cap using leather to prevent abrasion on a joint that is overly exposed.

Clothing Barrier	Affordance	Function	Feature	Application Example
Lack of protection against moisture.	Protection	Provides leak resistant or absorbent functionalities.	Choosing fabrics that have properties that are water resistive or absorbent.	Using a water-resistive or waterproof fabric for outerwear clothing to protect against weather alignments.
Uncomfortable designs that may cause health issues.	Comfort	Prevents pressure sores or overheating of body parts where fabric is gathered.	Seamless designs or wrinkle-resistant fabric that doesn't bunch easily.	Using draping techniques or silhouettes that use less seams that may prevent pressure sores.
Challenges with changing circulation or temperature.	Thermo-regulation	Provides the wearer control over temperature of the body.	Materials that provide circulation and moisture-wicking properties.	Using natural materials like wool, which have breathable qualities to prevent overheating for the wearer.
Difficult maintenance for cleanliness and for prolonging the life of clothing.	Antimicrobial	Provides ways clothing can fend off microorganisms and its ability to help prolong the life of a textile.	Materials that provide antimicrobial properties.	Using antimicrobial fabrics that can be made of a variety of textiles, such as polyester-vinyl or acrylics, in clothing that is frequently worn.
Lack of designs that conceal.	Discreetness	Allows the wearer to conceal unwanted clothing features or body.	Materials that camouflage or hide body and/or clothing features.	Designing looks that stylishly conceal a feeding tube that is connected to the body.

Clothing Barrier	Affordance	Function	Feature	Application Example
Lack of designs that communicate or signal to others.	Attention signaling	Enables the wearer to communicate to others to indicate presence or communicate a need.	Materials that reflect and have bold, contrasting colors.	Reflective tape on the seams of a material provides ways to signal to drivers at night that a person is present.
Challenges with clothing that does not fit the body tightly.	Compression	Provides the wearer psychological comfort and physical benefits that enhance performance and recovery.	Bonded fabric tape or highly durable yet stretchy materials that are tightly designed to hug the body.	Creating a garment that has little to no seams, such as a raglan shirt without shoulder seams, to create a fitted design.

Iterative Prototyping

The iterative process includes a variety of ways to approach design. Each iterative design helps measure the choice of materials, form, and usability, providing ways to better understand how fashion influences aesthetic preferences. In adaptive fashion, the iterative design helps prove a hypothesis. Is this sleeve silhouette providing great movement and comfort for the wearer? Does the fabric feel light when worn? The following are example questions that designers can reference for an iterative design process:

1. In what ways can we measure the frequency of use?
2. How can we assess and measure the importance of color?
3. How can we measure comfort?
4. How can we assess what material works?
5. What form and silhouette achieves the function we desire? How does it also align with the overall style and aesthetic of the garment?
6. Can we make quick and rapid samples of a garment to test for clothing closures?
7. How can we measure the weight of a garment based on the type of fabric?
8. How can we prioritize which measurements to assess and execute?

For Dorothy's jacket, a visual analysis and observation prioritized easier dressing elements like loops, a raglan sleeve, and a back pleat. Her preference for professional suit jackets and blazers was incorporated through these elements to create a functional yet stylish design. Figures 3.20–3.23 show each design function worn by model Karen Chan. After iterative prototyping, Dorothy's jacket featured the following adaptive elements:

1. Loops on jacket cuffs
   A sleeve was designed with two 7¼-inch (18.42 cm) zippers that opened into a 2-inch (5.08 cm) pleat to widen the sleeve hole. Additional elastic loops were added inside the hem of the sleeve to provide Dorothy extra leverage when pulling one side of the sleeve to throw over the other jacket sleeve. This allowed her to place both her arms into both sleeves quickly and easily with less pain.

2. Raglan sleeve for wider armhole
   A raglan sleeve was designed with ⅝-inch (1.59 cm) seam allowance. A cut was made into the neckband in two sections at the front raglan sleeve seam to allow for more arm movement.

Figure 3.21 Model Karen Chan wearing Access & Closure gray jacket design with a wider armhole. Photography by Alex Tosti. Courtesy of Alex Tosti.

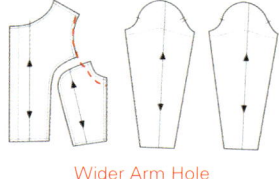

Wider Arm Hole

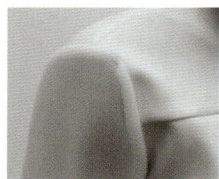

Figure 3.20a–b Model Karen Chan wearing Access & Closure black jacket design with embedded loops on the sleeve hem. Photography by Alex Tosti. Courtesy of Alex Tosti.

Figure 3.22 Sketch of the pattern design for the armhole for the white jacket by Grace Jun. Courtesy of Alex Tosti.

Approaches and Techniques

3. Back action pleat
   Adding an action pleat, or deep inverted pleat, in the back of the jacket for more freedom of movement in the upper body. The inverted pleat, including seam allowance, is cut from a similar or the same material of 3 inches (7.62 cm) in width, which provides a total extension of 6 inches (15.24 cm) when arms are raised horizontally.

Figure 3.23 Model Karen Chan wearing Access & Closure gray jacket with back pleat design. Photography by Alex Tosti. Courtesy of Alex Tosti.

Materials

Material exploration was one of the main factors that drove the project with Dorothy. The fabrics that provided the suit jacket structure did not have stretch. Therefore, each iterative design examined the combination of material and silhouette. Understanding the impact of such materials like neoprene or spandex was better informed by Dorothy's movement or range of motion in her upper body. For example, Dorothy moved her arms above her head only when she was not experiencing pain caused by her surgery. Some days, she was able to raise her arms, and on other days, she barely moved them above her waist. The motion of Dorothy's arms determined the sleeve design for the jacket. Range of motion (Chromat 2021) refers to how far a person can move or stretch a part of the body. Observations on body movement and dressing habits help designers understand a user's ability to reach out-of-reach objects.

Textile Considerations

Function	Behavior	Application Example
Durability	Synthetic fibers such as polyester can provide strength and fabric durability. It can produce materials that are more resistant to abrasion and superior in absorbency and comfort. A woven fabric stands up to the friction of braces and wheelchairs. While knitted fabrics that use highly durable fibers can be resistive, they can easily snag or be pulled when in contact.	Chapter 4, qXgo Applying an abrasion pad on the elbow area for wheelchair users. Reinforcement should be used at points of heavy wear and tear.
Easy maintenance	Water-repellent and shrinkage control finishes applied to natural fibers, such as cotton or wool, extend the life of a garment. Wrinkle-resistant polyester or nylon garments may be damaged from high laundry temperatures.	Chapter 4, LIULID Researching and obtaining performance materials that have finishes that protect against water or can withstand high temperatures.
Ease of dressing	Slippery materials like satin and silk provide easier ways for people to get dressed when the fabric is placed in the lining of a garment. Fabrics should also be lightweight so as not to cause fatigue when wearing. To avoid static electricity, certain synthetic fibers such as nylon and polyesters should be avoided.	Chapter 4, Midi-Rox Applying an interface or lining with a smooth fabric or using a double-sided fabric that already has a built-in material.
Skin irritation	People react differently to the feel of fabrics on their skin. Touch or tactile hypersensitivity may include skin irritation and other discomforts. Skin irritation may occur from fabric rubbing. Lightweight and soft fabrics with raised surfaces, such as cotton flannelette, may reduce the surface area in contact with the body. Personal preferences change for comfortable clothing as people age. Other irritations and discomfort occur due to the lack of compression from a material for people on the autism spectrum. Spandex and highly stretchy materials are recommended. Nylon and wool are hypoallergenic, which reduces allergy symptoms, and are also favorable choices.	Chapter 4, Ease Applying a sensory-friendly material that has the desired properties. Creating tag-less and seamless garments to avoid irritating the skin.

Approaches and Techniques 71

Body Measurements

Fit is an issue that has often concerned designers, apparel manufacturers, and consumers. The standardization of sizing systems is a necessity for the ready-to-wear industry (Burns 2011). Fit typically has two aspects—comfort and appearance. To better understand comfort, an analysis of body movements can be accomplished in the following ways: observation, photographic analysis, questionnaires, surveys, and interviews. A combination of these approaches is often advised to account for differentiation in body measurements.

Human movement provides both a constraint and a resource in the design of adaptive fashion. Movement influences pattern construction and also impacts size and fit. Pattern shapes need to correspond to the size and posture of the wearer. When a body moves against fabric, it can take extra energy to push against the fabric. For example, an individual's arm may not bend at a certain angle. If this is an issue, adding a permanent bend in the area of the garment can be achieved through patternmaking, such as the addition of pleats or the lengthening of fabric, which is advised to relieve some of the pressure. Bicycle shorts are made with a bend at the hip and a higher rise on the back to accommodate the rider's position. This would also apply to wheelchair users who are seated for long periods. The following list provides a way designers can prepare to document movement and measurement:

1. Username and date of observation
2. Description of clothing observed
3. Notes and fast sketches of the body in various positions (e.g., seated, standing, or hunched)
4. Description of body posture
5. Identifying parts of the body and their range of motion
6. Measurements from head to bottom
7. Sizes of bottom, top, and/or shoes
8. Circumference measurements of chest, waist, back, hips, ankle, upper arm, and upper leg
9. Length measurements for the entire arm, shoulder, inseam, outseam, upper and lower arm

Techniques

Designers employ skilled craftsmanship with a knowledge of design principles. Making clothing that is inclusive of disability needs must include fashion construction techniques, such as patternmaking and draping. The level of technique is not the focus of this chapter, but rather the way designers can approach maximizing an individual's needs. It is important to recognize the limitations of certain fabrics and materials to create better wearable designs.

Clothing Closures and Construction Techniques

From loops to buttons, clothing closures provide multiple entry points in clothing. Innovations like MagZip created by DNS Design, provide a quick, easy to snap zipper using well-positioned magnetic, making one-handed dressing easier. This closure would also benefit many people with limited dexterity. Yet not all closures need to be visible. For example, Chamiah Dewey Fashion is a UK clothing brand for people of short stature, measuring 4 feet 10 inches (147.32 cm) and under. Founder Chamiah Dewey learned about achondroplasia, a common form of dwarfism, during a collaborative youth program. While there were a few clothing options for people of short stature or little people, Chamiah noticed many were outdated, unflattering, and pricey. To address the lack of accessible clothing, the brand integrates adaptive features that are not visible or hidden, such as the Caitlin trench:

"We take into consideration the adaptations our consumer may need, such as discreetly hiding magnets with classic wooden buttons … All our products are designed for the short stature body and designed on a mannequin that is in the form of a woman with achondroplasia," says Chamiah (Dewey, pers. comm.).

Designed with pockets within reach for little people and belt loops purposely placed for fastening around the waist, the made-to-order Caitlin trench coat also has discrete magnetic fastenings to aid in easier dressing. The design is not only accessible but sustainable, made with Tencel Twill, a renewably sourced fiber. The trench fabric has the following properties: breathability, renewable, hypoallergenic, and anti-bacterial.

Thinking by making is reflected in the process of draping, patternmaking, or deconstructing garments. Construction and drape exercises provide ways to explore various contemporary and classic methods and techniques for garment design. For example, Richard Lindqvist's kinetic theory, which suggests live draping over the body, employs improvisation and critical thinking for designs to approach adaptive fashion solutions. The zero-waste design technique (patternmaking or draping) reduces textile waste going to landfill. In this technique, textile dimensions play a crucial role in zero-waste design and have long been used in the making of Japanese kimonos and Indian saris (Rissanen 2016).

Figure 3.24a–b The Caitlin trench coat, modeled by Robbie and photographed by Arch Photo Studio for Chamiah Dewey Fashion. Courtesy of Arch Photo Studio, Robbie.

Figure 3.25 Adaptive shirt designed with twisted drop sleeve by Ellen Fowles. Courtesy of Ellen Fowles.

Figure 3.26 Dress forms designed by SOUR in collaboration with Open Style Lab™ for standing, seated, and hunched body positions using laser cutting fabrication. Photography by Grace Jun for Universal Materiality Exhibition at Aronson Gallery in Parsons School of Design. Courtesy of Open Style Lab™.

Ellen Fowles, mentioned earlier in chapter 1, uses kinetic garment construction and ergonomic pattern cutting to enhance the functionality of the garments created in her capsule collection with Marian. Because the daily physical act of getting dressed was a constant barrier in Marian's clothing choices, the garment design above used this technique to shape a twisted drop sleeve with a zipper insert to widen the access points when donning or doffing the sweatshirt.

Techniques like this allow material tension and ease to become the visual and tactile apparatus for a made-to-measure process for both standing and seated positions. Through live draping, students and designers can comprehend gravity's effects on fabric and how to release tension

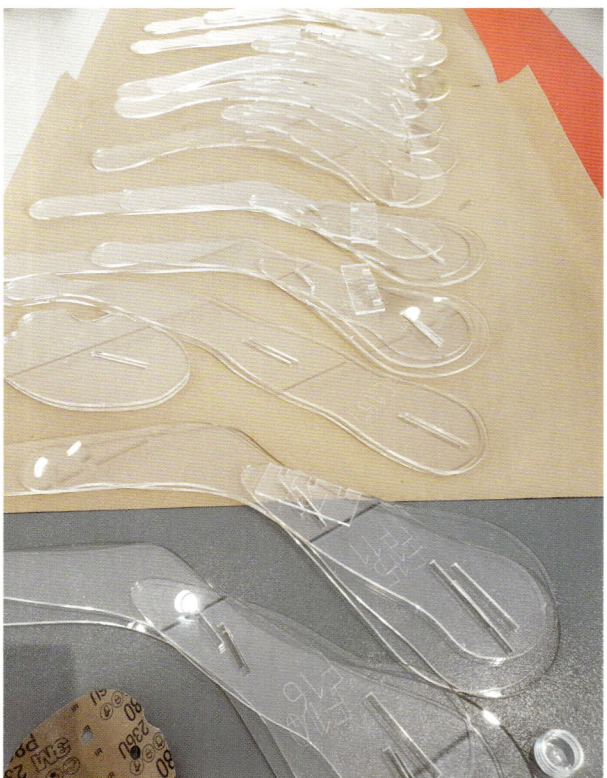

Figure 3.27 Dress forms pieces ready for assembly by SOUR. Courtesy of Open Style Lab™.

where needed for every body type and position, combining draping with classic metric patternmaking, using both flat and 3D construction systems. Commonly practiced in fashion education, students fit designs on professional models who have similar size measurements or **dress forms**—usually a size 6. To better understand a design in motion, fittings are conducted in the standing body and moving body positions. However, many dress forms are not designed to be inclusive of disability body types, nor are disability models commonly consulted to consider proportions and range of movement. While companies like SOUR have co-designed dress forms in the seated position into high-fidelity prototypes, many of the basic tools in fashion education have limitations in body forms that offer more ways to design with different body postures. SOUR is a hybrid design studio with the mission to address social and urban problems. In collaboration with a nonprofit called Open Style Lab™, the company creates inclusive body forms with different postures informed by wheelchair users, arm and leg amputees, and individuals with curvature in their upper back. Creating different body types with flexible motion and assembly is key to practical implementation in stores and exhibitions.

Alteration Instructions

From seated fit design to strategically placed closures, there are several ways to design and create alterations to make clothing more accessible. The following are commonly used alteration techniques to modify existing garments that better address the needs of PWD.

Loops

Loops provide a variety of ways for people to tuck, access, and leverage parts of clothing when getting dressed or taking off clothes. In adaptive fashion, loops can be designed as movement aids in the construction of trousers or jackets. This provides more areas to grip or hold while dressing. For example, pulling up long trousers or long sleeves may exert more energy and not be reachable for people with limited mobility. The following instructions are for creating a loop for the sleeves of a long-sleeved shirt, as seen in Dorothy's jackets.

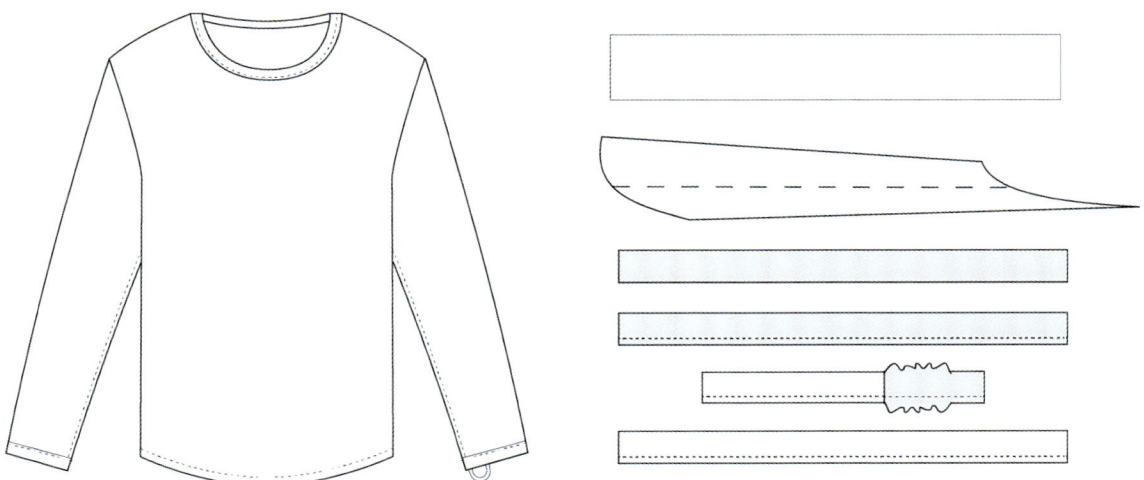

Figure 3.28a–b A shirt with a loop on the cuff of the sleeve and a series of folding steps for creating loops out of fabric cutout pieces.

Approaches and Techniques

Materials:
- Pick a long-sleeved cotton shirt. Knits or ribbed cuffs will not work well for this adaptation.
- Prepare 1½–2 in. (3.81–5.08 cm) fabric that matches the garment in color or style and has good elasticity or stretch. For example, shirring elastic is good for bunching fabric and easy to manipulate.

Instructions:
- Fold the fabric in half and stitch the ¼ in. from the edge of the end.
- Reverse the fabric inside out and make sure to flatten the material using an iron.
- To sew on the loop to the sleeves of the shirt, first, attach it along the seam allowance. Next, fold the fabric into a rectangular loop and reinforce the ends with another topstitch.

support when a person aids in transferring the wearer from bed to wheelchair. Denim or sturdy material is recommended for adding to the back of the pants. A long piece of fabric can be cut 1½–2⅛ inches (3.81–5.4 cm) wide. Fabric length is dependent on the weight distribution of the wearer and how much support is needed when pulling the back of the pants (The American Occupational Therapy Association 2020).

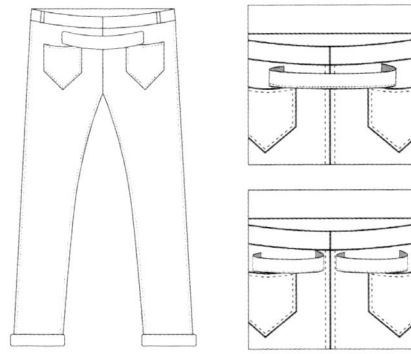

Figure 3.30 A pant design with two optional loops placed below the waistline: two back loops and one large loop.

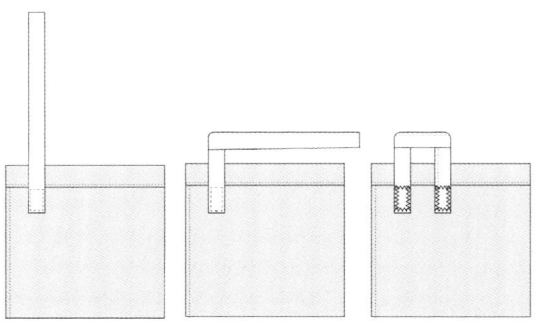

Figure 3.29 Three folding steps for the loop to be placed within a garment sleeve.

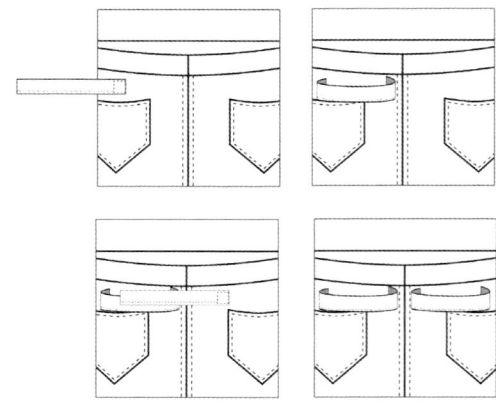

Figure 3.31 Steps to placing two loops on pants across the beltline.

Loops can also be placed on various parts of the pants for extra places to pull up or take off excess fabric when dressing. For example, a loop placed on the thigh area of the pants could be easier to reach than from the bottom hem. Loops are placed on the back of pants for extra

Magnets

Adaptive clothing for people with joint pain, arthritis, and limited hand mobility who may face difficulty with fastening and unfastening garments. Magnetic tape, snaps, and other fasteners can help with easier dressing. When sewing magnets into clothes, choosing one with a magnet cover is preferable. A cover helps to prevent the magnet from breaking or shattering. Designers should consider the type of magnet used to ensure it does not stick to other metal devices such as wheelchairs.

Materials:
- Magnetic snap (comes in two pieces or preferably covered by plastic pouch)
- Fabric or felt square scrap
- Fusible interfacing
- Seam ripper
- Fabric pen or pencil
- Plyer
- 1–1½ in. (2.54–3.81 cm) fabric squares or fusible square pieces (depending on your shirt buttonhole)

Note: A magnetic snap comes in two parts, the male (thinner side) and the female side (thicker side), and matching backing pieces (washers) that reinforce and close the snap onto the fabric. If using magnets with covered plastic pouches, stitching within the textured edges of the plastic pouch is required. Stitching too close to the magnet can break the seal.

Figure 3.32 A long-sleeve shirt with integrated magnetic closures.

Approaches and Techniques

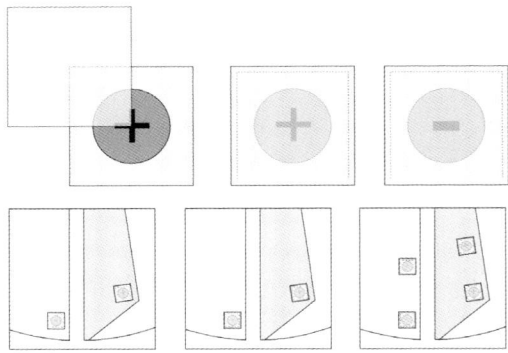

Figure 3.33 A positive and a negative magnetic closure placed on square fabric pieces.

Instructions:
1. First, create square fabric pieces in felt or fusible interfacing.
2. Mark the shirt where the prongs of the snap will be pushed through using a fabric pen.
3. Carefully cut these marks using a seam ripper and place the prongs through the shirt.
4. Place the female snap on the inside of the shirt.
5. Place the washer over the prongs and, using a plyer, fold your prongs in tight to the fabric.
6. Add the square fusible interfacing and fabric over the back of the magnetic snap.
7. Repeat these steps for the other magnets for the shirt.

Gusset

Gussets come in triangular or rhomboidal (diamond) shapes and can be found in many athletic wear shirts. These pieces of fabric are inserted into an existing garment to extend the range of motion (Chromat 2021) for a person who needs more flexibility. Gussets are traditionally added to the shoulders, underarms, or hems of shirts. For example, pocket gussets placed in the back of the armhole, like the Samsung Heartist examples in chapter 1, will increase movement for people using wheelchairs or crutches.

Figure 3.35 A diamond-shaped gusset pleat cutout for the underarm of a long-sleeve shirt.

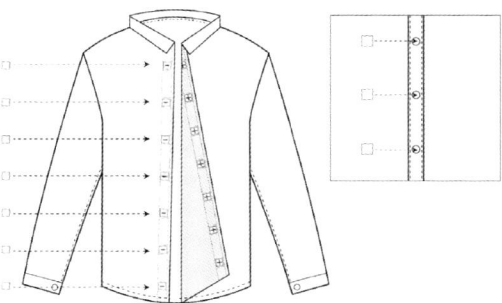

Figure 3.34 The left and right magnet placements within the shirt lining.

Materials:
- Pick a shirt or blouse with raglan or set-in sleeves.
- Prepare ¼ yard of fabric that matches the color or style of the garment.

Instructions:
1. Remove 2½–4 in. (6.35–10.16 cm) of stitching in the underarms of the garment.
2. Cut a 4–6 in. (10.16–15.24 cm) square fabric on the bias with a ⅝ in. (1.59 cm) inseam. You can topstitch for extra security.
3. Align the piece to the opening diamond shape.

Gussets that are designed from the start can reference figure 3.38. The center grainline of the sleeve should match to the shoulder seam. This gusset design runs through the underarm seamlessly and may provide comfort for people experiencing skin sensitivity, as seen in chapter 4, Ease. Alternative styles for increasing mobility in the arm area are petal and raglan sleeve designs.

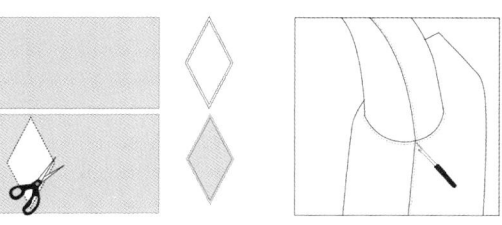

Figure 3.38 Two pattern pieces for joining the front and sleeve to create a longer built-in gusset style.

Figure 3.36 Two diamond patterns cut out from the fabric. The seams of the shirt's underarm are picked with a ripper to place diamond patterns.

Action Pleats

Action pleats are placed in the back of the shoulders to allow increased freedom of movement. This action pleat instruction depicts an example of the fabric running down the side seam of a shirt. Action pleats can extend from the shoulder to the sleeve or focus solely on the back shoulder for increased movement, providing greater stretch and movement when wearing non-stretchy garments.

4. Flip the garment inside out. Position the bias diamond square on the inside of the garment underneath the opening.
5. Press open seams and topstitch ⅛ in. (0.32 cm) from edges all around.

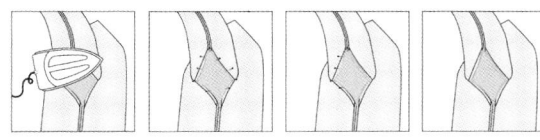

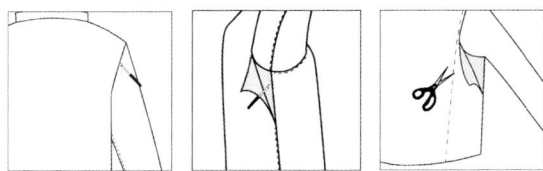

Figure 3.37 A shirt flipped inside out for positioning a diamond square fabric piece. The diamond square, or gusset, is steamed on top of the shirt.

Figure 3.39 The side seams of the sleeve picked with a seam ripper to insert an action pleat.

Approaches and Techniques

Materials:
- Shirt or dress with set-in sleeves
- 1/2 yds. of matching fabric color or style

Instructions:
1. Remove stitches behind set-in sleeves 1½–2 in. (3.81–5.08 cm) from the shoulder past the underarm seam.
2. Remove stitches in the bodice underarm and back waistline seams.
3. Draw a line parallel to the grainline from the armhole to below and then to the waistline seam.
4. Slash the bodice back apart on this line.
5. Trace the bodice on a separate piece of paper and add 3–4 in. (7.62–10.16 cm), increasing the width of the side bodice. This will make the top wider in the arm area.
6. Cut out a new pattern using this bodice with the extra increase.
7. Stitch the side bodice to the front arm at the underarm section and sew the edges to form a pleat.
8. Topstitch the bodice at the bottom of the plea to the other side to close at the waistline.

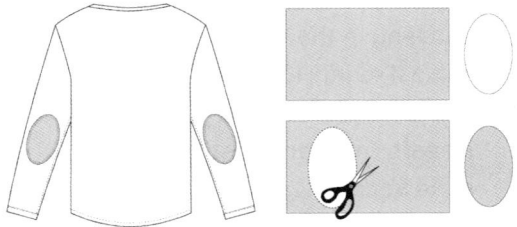

Figure 3.40 Two elbow patch pieces cut out to place on the back of a long-sleeved shirt.

Elbow Patch

Like the case study of qXgo in chapter 4, an elbow patch helps reduce the wear and tear of the sleeves in a garment. Materials like leather may present abrasion-resistant qualities for users repeatedly making movements that cause friction against assistive devices, such as wheelchairs or crutches. Damaged sweaters or tops with holes in the elbow area can be easily mended or redesigned with this patch reinforcement.

Materials:
- Pick a heavy yet flexible patch fabric like leather or corduroy.

Instructions:
1. Lay the top flat on a work surface with the damaged side up.
2. Measure the length and width of the hole.
3. Using pattern paper, sketch a curve connecting the endpoints of the length and width and mark a center point. An elbow patch may vary in size but typically can range from 4–6 in. (10.16–15.24 cm) long and 2–3½ in. (5.08–8.89 cm) wide.
4. Fold the paper in half on the straight line and cut out the pattern.

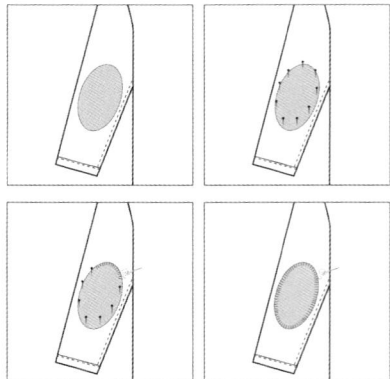

Figure 3.41 An oval-shaped patch pinned onto a sleeve that is flipped inside out. Blanket stitches connect the patch to the sleeve.

5. Flip the top inside out.
6. Pin your pattern piece onto your patch fabric.
7. Stitch around the edges of the patch.
8. Repeat on the other sleeve if you want symmetrical patches.

Ribbing Neckband

Rib knits are reversible double-faced fabrics. Ribbing fabric is a tight type of rib knit that is explicitly made for cuffs, waistbands, and neckbands. Adding a ribbed neckband to an existing garment may provide more stretch in the neck area and easier dressing—extremely useful during co-dressing situations, where a caretaker or assistant is helping to dress another individual. Furthermore, the ribbed texture may also help people who are blind or visually impaired to better locate and distinguish the neckline of a shirt.

Instructions:
1. First, measure the neckline of the existing garment and determine a loose or tight fit.
2. Matching or contrasting colors can be used.
3. Add 1 in. (2.54 cm) to the neckline measurement. The height or thickness of the fabric should be between 1–1½ in. (2.54–3.81 cm) without seam allowance.
4. Add ½ in. (1.27 cm) of seam allowance all the way around the fabric.
5. After your final measurements, trace them with chalk or fabric marker onto the ribbing fabric.
6. Cut two pieces of ribbing fabric with the stretch of the fabric going around the neckline.
7. Fold in half and stitch both ends of the fabric.
8. Flip the fabric and align it with the inside of the shirt neckline.
9. Topstitch the seam allowance down to the neckline side or finish the seam allowance with a zigzag stitch.

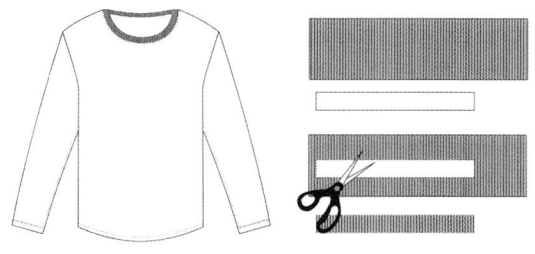

Figure 3.42 A cutout collar of a ribbed material to create a stretchy neckband.

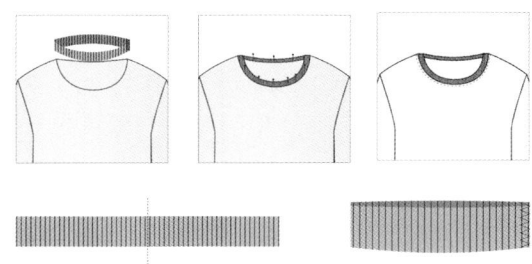

Figure 3.43 A folded ribbed fabric prepared to be stitched with a zigzag foot 1/4 in. (0.64 cm) inseam from the raw edge. The piece is then pinned to the neckline to be assembled.

Materials:
- Ribbing fabric

Note: Ribbing is usually sold by the yard or by inch. It may be purchased on a bolt just like regular fabric or in a tube with no cut edge.

Front Pockets

Pockets can be placed on different parts of a pant design. Typically, sewing front hip pockets or on-seam pockets are constructed while making the rest of the pants. Both pockets are integrated with lining. Adding additional pockets, such as kangaroo pouch shaped pockets or pocket squares, can be useful for users needing to carry more things closer to the body. For example, many wheelchair users may drop items or have difficulty accessing pockets on the side of pants. Front pockets designed by the thigh area are accessible while seating.

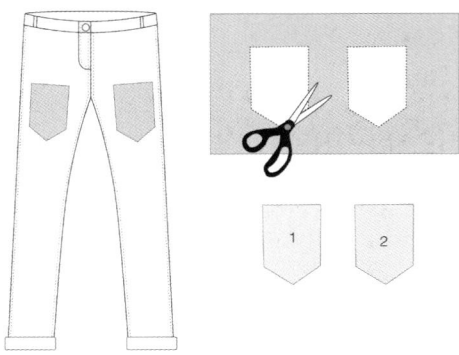

Figure 3.44 Two cutout pocket pieces with ¼ in. (0.64 cm) seam allowance.

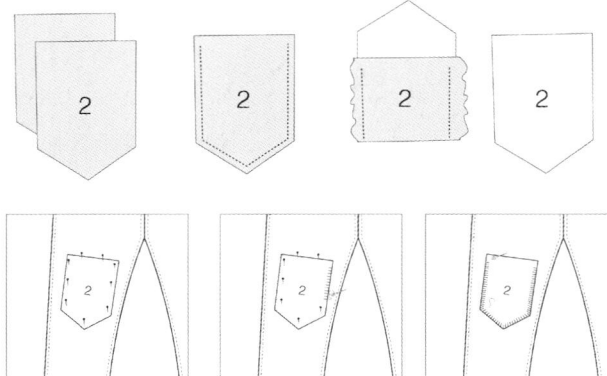

Figure 3.45 Close-up of the second pocket cutout and flipped inside out to be placed onto the front part of the pants. A blanket stitch is recommended for security or extra texture to better locate the pocket for those who are blind.

Materials:
- Choice of cotton, denim, or other material for pocket design.

Instructions:
1. Cut out four pocket shapes and stitch two together.
2. Flip inside out and create a ¼ in. (0.64 cm) seam on the pocket.
3. Use a serging or zigzag stitch, creating a clean finish on the raw edges of the pocket.
4. Fold top edge under ½ in. (1.27 cm) and stitch.
5. Fold side edges under ¼ in. (0.64 cm) and press.
6. Fold bottom-angled edges under ¼ in. (0.64 cm) and press.
7. Sew to the front of the pants. Start at the top and sew all the way around. Leave the top open and backstitch at both ends.

Pant Zipper

Inserting an invisible zipper in the inseam of pants may facilitate dressing for a person wearing a cast, brace, or incontinence appliance. Positioning the zipper on the seam or even toward the front of a pant may provide easier reach for wheelchair users. Zippers can also be adorned with tassels that help a wearer reach and easily pull the zipper off.

Materials:
- Invisible zipper
- Zipper foot
- Thread
- Scissors
- Pins

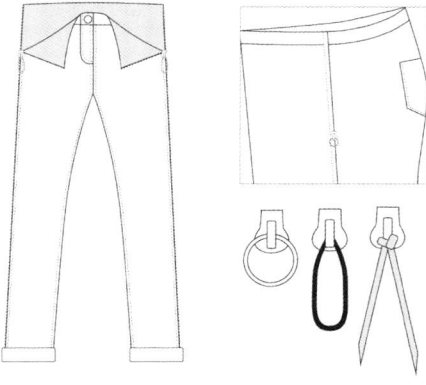

Figure 3.46 Various attachments that help extend and provide accessible grips for a pant zipper.

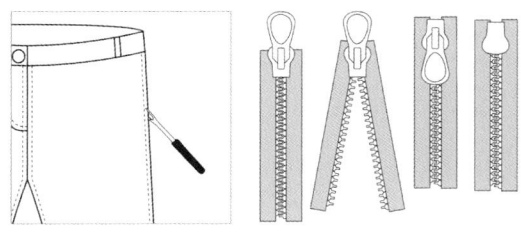

Figure 3.47 Creating an incision for side zipper placement on the side seam.

Instructions:
1. Choose a zipper length to create an opening on the side of the pants.
2. Remove stitching of the pants inseam to match the desired zipper length.
3. Turn pants inside out and pin the zipper down where the opening is.
4. An inside reinforcement patch can be sewn into the garment to protect the fabric against abrasion. Iron on a patch to protect the pant fabric and stitch around the edges of the patch for extra durability.
5. To protect the skin from the abrasive action of the zipper, insert another fabric piece.
6. Fold a strip of fabric and stitch across the long ends.
7. Stitch the piece of fabric underneath the zipper on one side, flip it over, and stitch the other side while open.
8. Flip the pants inside out.

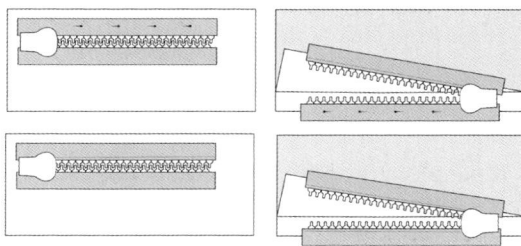

Figure 3.48 The placement of an invisible zipper with extra fabric on the inside of the pant side seam.

Pant Extension for Seated Body

The fit and style of pants often don't accommodate the sedentary position or seated body. Pants tend to crease with wrinkles, making the wearer look unattractive, and can be uncomfortable when sitting for a prolonged period. The pants' back length is increased because the body is curved when seated, while the front length is decreased to prevent bunching of fabric in the abdomen area. A higher back length may also prevent the wearer from slipping forward and exposing the back. Adding elastic fabric where parts of the body are bent, such as the elbows or knees, provides comfort and easier donning and doffing.

Approaches and Techniques

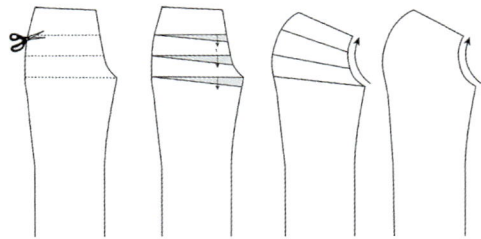

Figure 3.49 Slashing the original design to create extra room in the front area of the pants.

Materials:
- Prepare long trousers, preferably one size up
- Scraps of paper and tape

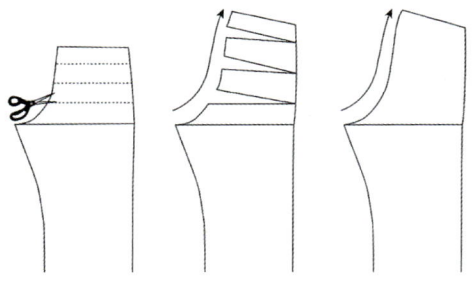

Figure 3.50 Extending the pants for extra room on the back using the slashing technique.

Instructions:
1. To add extra length in the waistband, mark the hip rise line along the grainline.
2. Draw one or two parallel lines dividing the pattern above the hip rise line into thirds.
3. If only slashing once for a slightly curved back, 6–10 in. (15.24–25.4 cm) above the hip rise line is enough.
4. Create one or more slashes from the center front to meet the hip rise line to create a more curved look around the crotch and side seam area.
5. Spread the slashes 1½–2 in. (3.81–5.08 cm) or as much need for the body type.
6. Redraw the center back and side seams.
7. The center front should be shortened or eliminated, so shift waistline darts towered the side seam. The front seam can be shortened by taking in 4 in. (10.16 cm), smoothing out the front part of the pants.

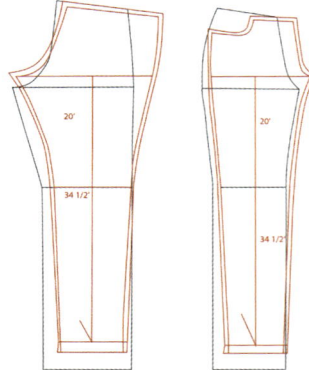

Figure 3.51 An overlay of pant patterns that depicts the difference between original pants and pants that have used the slash technique to extend the top front and back area of the waistline.

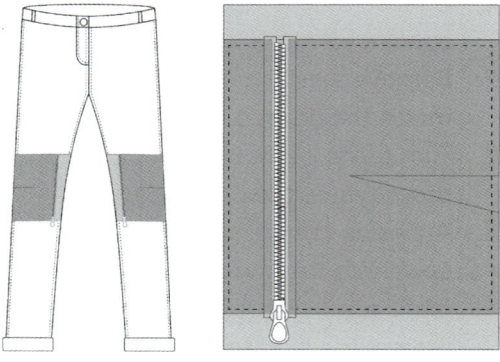

Figure 3.52 Placement of a dart and zipper near the knee joint, providing increased movement for the knee joint. Zippers can perform as hidden movement aids, expanding the fabric by the knee.

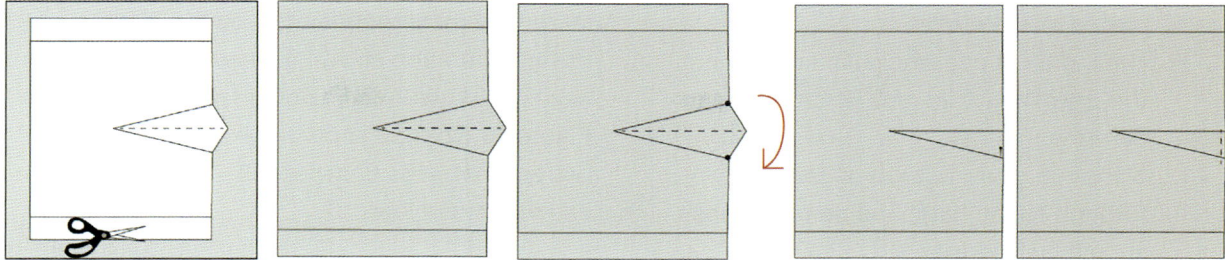

Figure 3.53 Process of creating a dart on fabric to alter pant design.

Sleeve Design for Seated Body

The bodice or front of a garment's length will need alteration when designing for the seated body. The front will decrease in length, and the back will increase with an extended collar. Adjustments to the bodice will vary depending on the wearer's body shape and preference, but sleeves must be altered for a bent arm position.

Materials:
- A prepared sleeve from a jacket or top of your choice
- Scraps of paper and tape

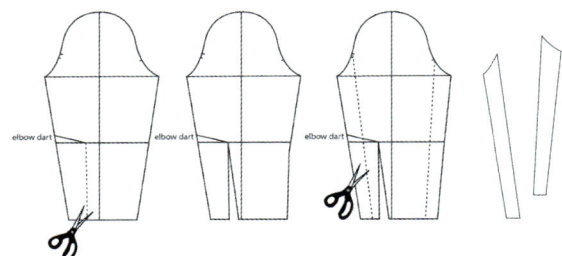

Figure 3.54 Steps in creating a sleeve pattern by altering the elbow area.

Instructions:
1. To create a two-piece sleeve, first identify or create an elbow dart.
2. Slash through the elbow dart to the wrist.
3. Fold the dart and spread the slashed portion.
4. Place paper underneath the slashed portion and tape in place.
5. Open the dart and cut the sleeve at the seamline.
6. Join the front and back underarm to create matching pieces.
7. Cut the front to spread open the dart.
8. Redraw the elbow dart and extend through the underarm sleeve. This will create a forward-slanted look.
9. Redraw the front and back seam of the upper sleeve.

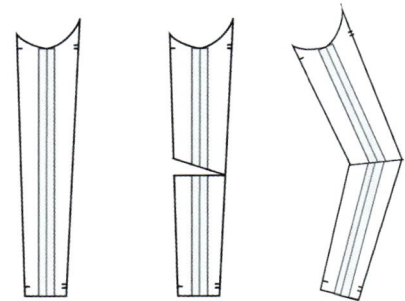

Figure 3.55 A fabric piece of the sleeve that is reconfigured to a different bent shape. A single mark indicates the front, and the double slash lines indicate the back.

Approaches and Techniques

10. Add ⅝ in. (1.59 cm) seam allowance. There should be two distinct pattern pieces of the sleeve.
11. Next, draw a perpendicular line against the grainline (⅕ in. or 0.51 cm) on both patterns.
12. Slash the seam but not all the way through to the back. The pattern pieces should curve forward.

Like the use of darts to extend pants, a pleat can be placed at the elbow area of a sleeve to provide increased movement and flexibility. Unlike the design alteration above, a sleeve pleat provides flexibility when extending and retracting the arm. Figures 3.57–3.58 show an example of a sleeve pattern that uses 6 inches (15.24 cm) for creating an extension pleat.

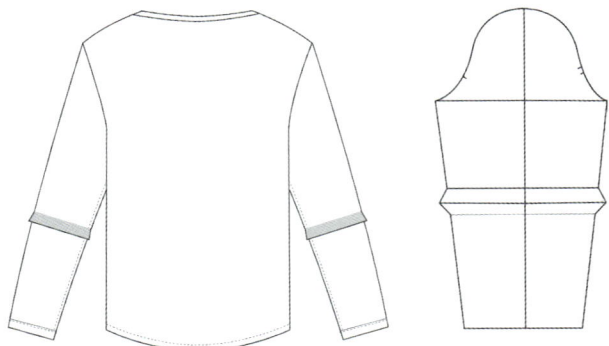

Figure 3.57 The back of a shirt with a sleeve pattern on the right using an elbow pleat.

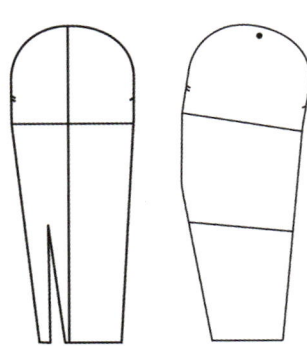

Figure 3.56 Movement of top shoulder mark to extend elbow area of the sleeve.

13. Move the top of the shoulder to the front by marking a new center (¾ in. or 1.91 cm), causing the sleeve to be positioned more forward.
14. Cut out fabric on both pattern pieces and construct them together.

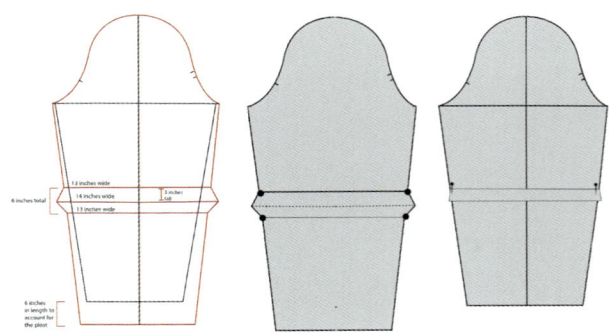

Figure 3.58 Steps to create 6 in. (15.24 cm) pleat fold.

Conclusion and Design Evaluation

All tests must end when iterative prototyping no longer helps further progress. An assessment of measures and methods used helps evaluate the impact of the design. How does the final design address the initial design requirements? Does the final design meet the expectations of the wearer? Was the design created for an individual bespoke use case or a broader user base?

First, collect and analyze current data and final design iterations. Next, create an Inquiry Guide that helps examine the design in more detail by organizing three categories to form an evaluation. The chart below was used to discuss the details of the final design with Dorothy.

Inquiry Guide

Category	Question	Discussion	Object in Question (e.g., pictures, a prototype, or anything visual)
Observation	What part of the jacket design do you want to examine?	Closures are important for getting dressed.	A muslin prototype with magnetic closures.
Interpretation	Why are there magnetic buttons on the jacket?	Magnetics are one type of clothing closure that can help with easier dressing.	The placement and number of magnetic closures on the jacket.
Evaluation	Do you want to keep these specific magnetic buttons?	I think the magnetics could be smaller. It will weigh less and, therefore, not tug on the fabric.	Magnet size and weight.

Then, develop a rubric of evaluative approaches and interview the wearer individually and in a group. Finally, create a scoring standard that connects to the design requirements to critique the final design elements.

Figure 3.59 Dorothy Jones, wearing a gray jacket from Access & Closure collection. The jacket was designed with a wide armhole and smaller magnetic closures after evaluation. Courtesy of Grace Jun.

Design Rubric

Topic	Critique	Example Questions
Personal connection	Is there a meaningful personal connection between Dorothy and the final design?	Does it incorporate Dorothy's style, identity, and preferences in color, form, and material?
Comprehensive observation	Was there enough observational data collected throughout the process?	How does observing Dorothy's dressing behaviors inform the closures and silhouette design choices for the jacket?
Visual analysis	How was Dorothy's style incorporated into the design?	Does the jacket aesthetic match Dorothy's existing wardrobe and looks?
Functional performance	What function did the jacket have to perform?	What functional priority did Dorothy place when constructing the prototypes for the jacket design?
Context	What are the different situations in which Dorothy wanted to wear this jacket?	How were the environment and context considered in the material choices of the jacket?
Informed judgment	In what ways did Dorothy decide on the next steps of the jacket design?	How were the design constraints tested, and what were the takeaway factors of each iteration?

In conclusion, design links theory and practice, elegantly managing the complexity of open-ended problems found, such as fashion design solutions for disability. Designers create processes to deal with such specifications while negotiating between what is known in fashion techniques and the particulars of an individual situation. Design knowledge may lean less toward finding answers and more toward methods leading to answers. Successful design is iterative, helped by feedback in the process. This feedback is found in actual experience. Adaptive fashion designs work when the process centers on the experiences of PWD.

References

Burns, Leslie Davis, and Kathy K. Mullet. 2011. *The Business of Fashion: Designing, Manufacturing and Marketing*, 4th ed. London: Bloomsbury Publishing.

Chromat. 2021. "Chromat x Tourmaline SS22." Blog. Last modified September 15, 2021. https://chromat.co/blogs/news/ss22 (accessed August 14, 2023).

Gupta, Deepti. 2011. "Functional Clothing—Definition and Classification." *Indian Journal of Fibre & Textile Research* 36 (4): 321–326

Hamraie, Aimi. 2017. *Building Access: Universal Design and the Politics of Disability*. Minneapolis: University of Minnesota Press.

Ladau, Emily. 2021. *Demystifying Disability: What to Know, What to Say, and How to Be an Ally*. New York City: Penguin Random House.

Rissanen, Timo, and Holly McQuillan. 2016. *Zero Waste Fashion Design*. London: Bloomsbury Publishing.

The American Occupational Therapy Association, 2020. "Occupational Therapy Practice Framework: Domain and Process Fourth Edition." *American Journal of Occupational Therapy* 74 (2): 1–87. doi:10.5014/ajot.2020.74S2001 (accessed August 14, 2023).

Chapter 4
Case Studies, Stories, and Interviews

The personal experiences of real people are essential to an inclusive design process, especially for adaptive fashion. This chapter explores different ways to design with PWD. The following stories provide design educators, practitioners, PWD, and students who want to take a collaborative approach to designing adaptive fashion with an inquisitive lens. Rather than identifying disability as a design problem paradigm, these examples demonstrate how centering on identity-based stories resists normalization and conformity. These stories also depict how many creative solutions have already been explored by PWD. PWD can adapt and introduce innovative approaches to design—especially fashion. Each section reveals how collaborative teams started to explore adaptive fashion, created unique plans, managed roles within a diverse team, and reflected on their progress. Many of the design processes are biased toward field-based studies using subjective measures. Some groups used objective measures like laboratory-based evaluations, such as tools to measure water-resistant textiles, to evaluate material impact in clothing design. Overall, all stories demonstrate the complexity of design—multidisciplinary collaborations and identifying accessible or adaptive features in fashion.

This chapter asks if co-design, applied during the ideation stage of an adaptive fashion process, can create more innovative outcomes. Yet collaborations require multiple stakeholders to cooperate because co-design processes use team approaches involving people from different backgrounds (Mahr 2014). It requires an investment of personal time, initiative, a specific skill set, and the ability to adapt or change perspectives within a team. While each design collaboration focuses on the needs and desires of one person with a disability, the process offers students ways to approach inclusive design.

Processes that are inclusive of disability build interpersonal communication skills, accountability, and a space for people to share other ways of thinking and processing. Most importantly, they give educators and students an opportunity to learn about the inaccessible barriers faced by the disability community as creative opportunities.

There is no singular disability experience as depicted by scholars and writers in the Further Reading section of this book. Therefore, the following narratives demonstrate a multitude of perspectives and approaches toward adaptive fashion. Stakeholders, collaborators, materials, different types of disabilities, different skills and backgrounds, personal experiences, and more have shaped each design process. The narratives explore how each team negotiates conflict and adapts to unplanned experiences when working as a multidisciplinary team. Each story has been portrayed with permission from the participants and teams I've encountered throughout my time teaching at Parsons School of Design and contributing to a nonprofit called Open Style Lab™. These design collaborations occurred between 2014 and 2021 and, therefore, are described for that moment. Each person living with a disability has chosen self-descriptive words that best depict themselves for each case study during the time of collaboration. While many of the people and designs may have changed, the stories were depicted in those moments to capture the design process as it were. Each story is also presented in order of design complexity and topic. For example, the first three case studies (Unparalleled, Midi-Rox, and The Little Black Bag) demonstrate processes that stress disability narrative heavily in research. Furthermore, the making process of each design uses primarily fashion techniques, whereas the last case study examples integrate complex technologies and other tools.

The focus of this chapter is not on age or disability, but on co-design approaches applied to a variety of adaptive fashion outcomes. Teams of designers, engineers, and therapists with and without disabilities are each portrayed with a design process and end product. Interviews with select people were chosen to bring more attention to a particular topic, such as inclusive representation or the complexities of textiles. I am extremely grateful to receive the trust and permission to retell each project from everyone featured in this chapter. In doing so, I hope to convey the nuances unique to each project while bringing a cohesive narrative for students to refer to when needed. Each collaboration serves as a case study example to reference techniques and types of wearable design outcomes. The case studies demonstrate how each team utilized co-design processes, human-centric design approaches, a thoughtful investigation of adaptive fashion starting with a person with a disability, and the willingness to share what was learned.

Design Case Studies

Interview: Inclusive Representation—Christina Mallon

1. Unparalleled
2. Midi-Rox
3. The Little Black Bag

Interview: Understanding the Use of Materials—Angela Domsitz Jabara

4. Trans-Skirt
5. qXgo
6. Swipe

Interview: Human Factors and Occupational Therapy—Michael Tranquilli

7. Zipback Jacket
8. Ease
9. Modiste
10. Revolve

Interview: Interactive Garments and Textiles—Dr. Jeanne Tan

11. Versa Vest
12. Avisly
13. LIULID
14. Warmed Bomber

Interview
Inclusive Representation—Christina Mallon

Christina Mallon is the Director of Inclusive Design at Microsoft and Board Member at Open Style Lab™. She is a woman at the forefront of an important movement toward inclusivity in design and advertising. She has a unique and special voice, championing individuals often ignored by these industries—particularly those with disabilities. Christina's work around inclusive design has received worldwide attention and she was awarded LinkedIn Top Voice in Design 2020, Ad Age 40 under 40, and Business Insider's Rising Stars.

Grace
What has led you to do the work that you're doing today?

Christina
I am the Director of Inclusive Design at Microsoft. I became an inclusive designer because about eleven years ago, my arms slowly became paralyzed. I realized that there were little to no solutions for disabled people and that the world was not designed with disability in mind, which pushed me to join Open Style Lab™. I was enlightened about how to use design, engineering, and occupational therapy through cross-collaboration to be able to create better designs. I focused my career on these collaborations moving forward, and that brought me to Microsoft.

Grace
You have such an important job, so how important is getting dressed for this role?

Christina
From a larger standpoint, dress is important because you legally must be dressed in the United States. It's required to live and work in society. Dress not only has to cover your body but also serves the purpose of expressing your personal style. This is especially important to designers as it's another creative outlet for them. How we represent ourselves through clothing is very important. When it comes to the actual role, I think style and being able to interact and use products and services that are beautiful and accessible allows one to feel accepted in the world. So style, in particular, is very important for assimilation and self-acceptance. That's why that's a great focus on the products created at Microsoft.

Grace
Was there anything that you found inaccessible and frustrating to wear?

Christina
As a New Yorker, you're wearing a coat five months out of the year. Not being able to put that on yourself for protection from the cold of New York was one of the biggest pain points. That's why I sought out Open Style Lab™ because they could help me with that solution. This was not just for a functional need but from a self-expression standpoint.

Figure 4.1 Christina Mallon working with accessible technology at her desk. Courtesy of Christina Mallon.

Grace
Could you describe what that process was like, the teamwork, or the people you met?

Christina
It was quite an amazing and enlightening process that I wish every disabled person could experience if they wanted. Companies sometimes just think only about function, function, function! The great thing about Open Style Lab™ is how the organization thinks about function and fashion on the same level. The team at Open Style Lab™ is one that is cross-collaborative and sees the subject-matter experts as co-collaborators, which gave me the empowerment to speak up and to provide insight into my lived experience. Many times, disability is presented as a charity, but Open Style Lab™ uses the social model of disability, and the experience that I had was that you have this lived experience that is valued as expertise.

Companies are only going to survive if they innovate, and the disabled community is something that everyone will be a part of at some point in their lives, so including people with disabilities is crucial to innovation. This thought pushed me to get involved in inclusive design and to bring that to large organizations.

Grace
You mentioned a lot about the process and how collaborative the Open Style Lab™ course was at Parsons. Students could come visit you at home and understand your daily living. Is there a specific moment you recall that exemplified collaboration?

Christina
I had a wonderful experience with the students at my home. Every single time that you can't do something, it's painful and sometimes hard to describe. The home visits allowed them to visually see the challenges that I run into day-to-day. Co-design of accessible solutions is not

Figure 4.2 Christina Mallon using her feet to pull down a cell phone holder to type with her toes.
Courtesy of Christina Mallon.

the norm. Usually, a nondisabled designer uses their pre-consisting notions around disability to create accessible goods that end up not working for the intended audience. The core of Open Style Lab™ is co-design.

Grace
What is an inclusive designer? Do you see more roles like this today?

Christina
When I think of design, anyone is a designer if they make decisions that affect someone else because they're designing the world. An inclusive designer specifically identifies how the lived experience and bias, consciously or subconsciously, are reflected in a person's decision-making process. The designer is part researcher and part organizer, helping technical designers go through the process of ensuring that they have the knowledge to stop biases when creating. They can ensure that co-design is from the beginning to the end, from ideation to solution. That ensures that a product works for more people.

Grace
What advice would you give to designers, creatives, and fashion students (with or without disabilities) today about getting into inclusive design?

Christina
My biggest advice is to really spend time researching. Who is greatly ignored? How can we study design history, the biggest designers now, and think about other fields? How can you avoid bad design or inaccessible design if you just really look outside your current world? We must have multidisciplinary education and teammates with diverse lived experiences. I think that's key if you want to have a future in inclusive design.

Grace
You hinted at disabled creatives and that having disabled creative leaders is essential to inclusion. Could you tell us how important the lived experiences of disabled people contribute to creativity?

Christina
As August de los Reyes stated, disabled people live in a world not designed for them, which challenges them to think and come up with creative solutions all the time. It's beneficial for design organizations to have disabled people in leadership roles to increase innovation.

Grace
Terms like inclusive design, universal design, and accessibility all ask to make something better for more people. How do you see this evolving in the future?

Christina
Many practitioners in these fields are starting to integrate intersectionality into their work. Kimberlé Crenshaw coined the term intersectionality, which is simply about how certain aspects of who you are will increase your access to the good things or your exposure to the bad things in life. I'm seeing a lot of inclusive design practitioners, and bigger corporations are starting to value inclusive design, which allows for better products. These are the two ways I see design changing in the future.

Conversation Points
- How do you define inclusive design in your projects?
- How can you conduct desk research to avoid creating inaccessible moments during your design process?
- What is being ignored or looked over in the process?
- How does the environment impact your design process?
- Have you assessed your own biases and assumptions about disability?

Case Studies
Unparalleled
Christina Mallon, Claudia Poh, Julia Liao, Estee Bruno

Figure 4.3a–b Christina Mallon wearing a black coat design called Unparalleled. Courtesy of Open Style Lab.

Key Functional Features
Breathability
Weight
Easy to don and doff

Summary
Unparalleled is a stylish two-piece black coat with an accent belt that can be worn hands-free.

Design Goal
To create a coat that could be worn and taken off independently in public settings.

Team
Christina Mallon: Client, subject-matter expert, and inclusive designer
Claudia Poh: Fashion designer
Julia Liao: Product designer
Estee Bruno: Design technologist

Background

Whether it's occupation-specific (like a nurse's scrubs or a firefighter's protective suit) or merely a means to look professional, clothes play a big part in where we work and how comfortable we feel in our environment. Most workplaces require a dress code to create a sense of comradery among their employees. It displays unity and mutual understanding not unlike that of a school uniform, which aims to instill equality by removing the focus on possessions and appearance. One of the most common garments worn to work are coats, which are both stylish staples and protective pieces from weather conditions. This intersectionality of fashion and sportswear has had a long history, tracing from designers to specific material innovations, such as synthetic textile development in the 1960s and its influence on the space-inspired looks created by André Courrèges. Clothes are essential to the physical, social, and psychological well-being of everyone, which is why independent dressing boosts the confidence and self-esteem of PWD. The following coat design process explores a possible independent dressing solution for people living with disabilities who commute outdoors.

Introduction

A full-time advertising executive, Christina also has a rare form of ALS that has left her arms and hands completely paralyzed. Over six years, she progressively lost the use of her upper extremities, making dressing and traveling difficult. Her active lifestyle and daily commute to work were some of the most important concerns when dressing every morning. When members of the team first met Christina, she shared some of her goals, which included being able to don on and doff her winter coat more easily. She didn't want an assistive device to support her wearing a coat.

Design Process
1. Observation
2. Goal Setting
3. Prototype Iterations

Observation

"One of the main things I was hoping for by working with Open Style Lab™ was to be able to have a coat that I could put on myself before riding the subway."

Christina Mallon (pers. comm.)

Christina lives by herself in a Tribeca apartment and has a nurse who stops by every morning to help her with daily tasks before commuting to work. When the team first went over to observe her environment and everyday routines, they noticed assistive devices throughout her home. A few of them were upside down hooks on the walls—designed by occupational therapists at NYU—to help her don and doff clothing. She pointed out that while they

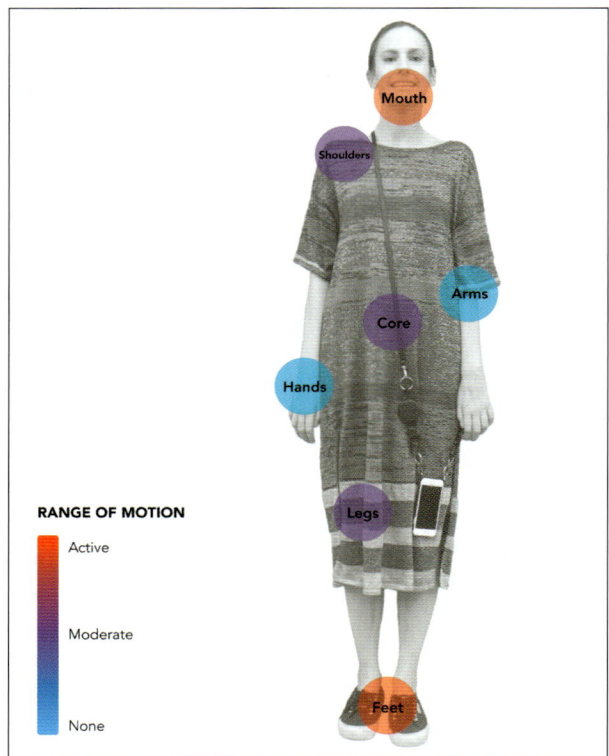

Figure 4.4 Sketch of Christina Mallon and observational notes on her range of motion by Claudia Poh, Julia Liao, and Estee Bruno. Courtesy of Claudia Poh.

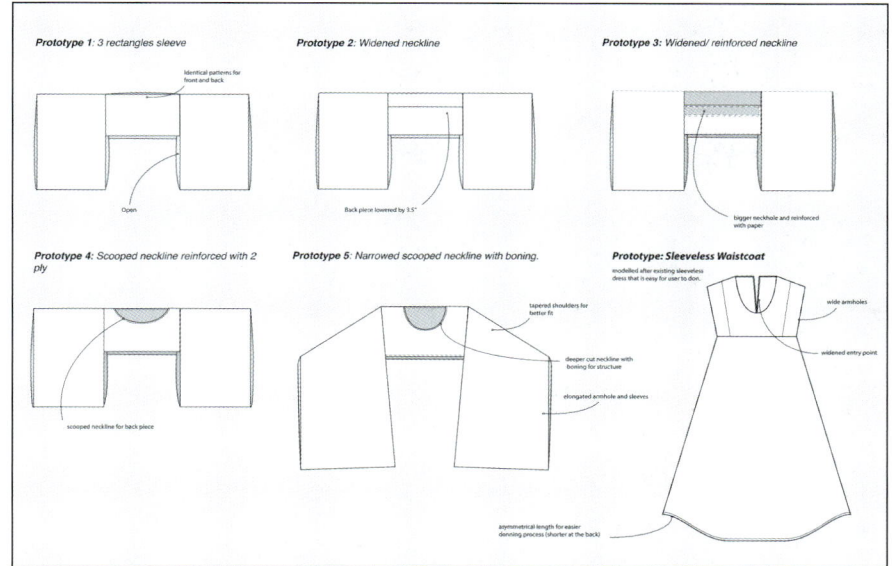

Figure 4.5 Sketch the coat's design iterations by Claudia Poh. Courtesy of Claudia Poh.

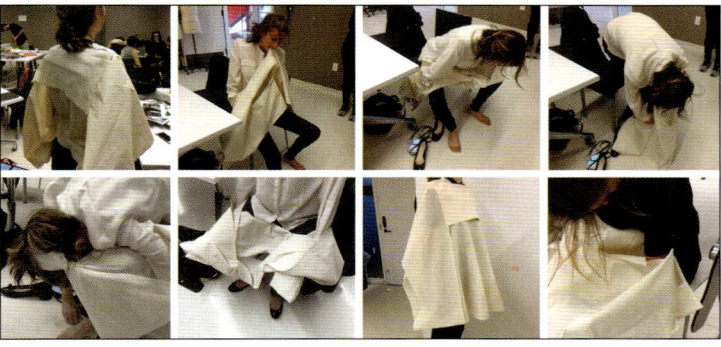

Figure 4.6 Photo of Christina Mallon trying on multiple design iterations for best fit and donning and doffing experiences. Courtesy of Claudia Poh.

worked well with stretchy fabrics and heavier materials, they didn't enable her to dress independently. Therefore, she always had to ask her friends or nurse for assistance, which took even longer in the colder months. Lastly, the team learned that Christina engages with clothing using various parts of her body—for example, by using her mouth or shifting her upper body weight. So, they created a sketch to help identify Christina's range of motion when she put on clothes (see fig. 4.4).

Goal Setting

The team focused on design solutions that were able to (1) reduce the amount of time it took Christina to put on her coat at home and (2) help her independently don and doff the garment in any setting (The American Occupational Therapy Association 2020). They also identified four necessary design attributes:

1. Warm
2. Comfortable
3. Flexible
4. Insulated

Prototype Iteration 1

Christina's closet is filled with colorful patterns from brands like Club Monaco and Zara. She prefers garments that are baggy or loose (like a sleeveless dress) so that she can quickly and comfortably undress herself. When the team looked at her wardrobe, they found a poncho she no longer wore because it was difficult to put it on.

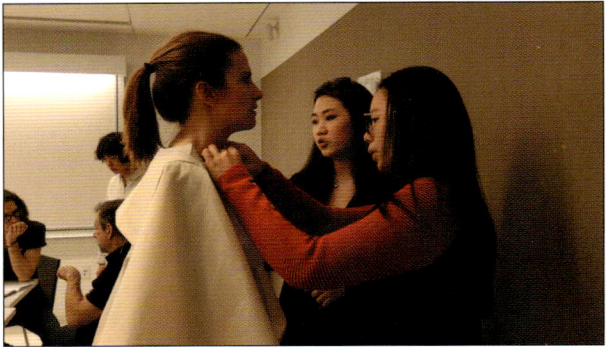

Figure 4.7 Photo of Christina Mallon with Claudia Poh and Julia Liao, making last-minute changes to the neckline of the final coat iteration.
Courtesy of Estee Bruno.

Figure 4.8 Photo of Christina Mallon wearing Unparalleled, a two-piece coat design paired with denim pants.
Courtesy of Christina Mallon and Claudia Poh.

Designed in a heavy fabric with intricate side openings, Christina used to rely on the corner of her bedpost to put it on. However, when the sides of the garment detached, it became impossible for her to hook it in place while she tried to poke her head through the neck hole. Based on this poncho and her collection of sleeveless dresses, the team was inspired to explore the construction of kimonos to create a two-piece winter coat.

After observing Christina put on coats with long sleeves, the team members realized a garment with a stable and wider round neck opening would be most useful. Their first prototype featured boning and stretch fabric, so it wouldn't need any closures or other fasteners that were difficult to reach. By inserting her head through the round neck hole, Christina was able to flip the garment by standing up before gravity pulled it down around her shoulders. And the larger armholes let her easily slide her limbs into the coat once it was over her head.

Prototype Iteration 2
The team continued developing a sleeveless dress-coat after examining Christina in the first prototype. For their subsequent iterations, they merged the styles of a kimono and sleeveless dress using muslin and added a high waistline and round neckline. What resulted was a garment that took Christina less time to put on herself than any of her existing coats.

Prototype Iteration 3
For the final prototype, the team focused on the coat details and materials. They created a waistcoat design with a silky interior lining that let Christina slide into the garment faster and easier and included a warm but lightweight exterior fabric accentuated by a belt for a fitted, feminine touch.

Material Studies
The final material chosen for the coat was a light wool fabric that was stylish but also warm enough for the winter.

Materials
- 100% wool felt
- Jersey
- Cotton

Final Design Outcome
The final design is a two-piece coat with an adjustable belt. It features a ribbed stretchy material in the back of the neck opening and is reinforced with boning to keep it sturdy while letting Christina slide her head through it like a sweater.

Midi-Rox
Roxinne (Roxy) Gaussite, Michael Tranquilli, Alyssa Wardrop, Ray Luo

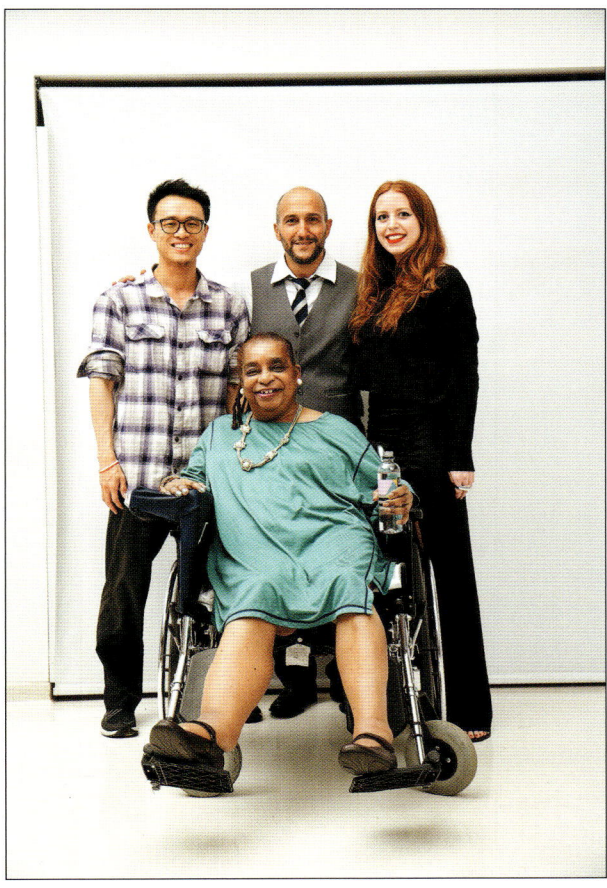

Figure 4.9 Photo of Michael Tranquilli, Alyssa Wardrop, Ray Luo, and Roxinne (Roxy) Gaussite, who is wearing the Midi-Rox green dress. Photography by Kilian Son for Open Style Lab. Courtesy of Open Style Lab.

Key Functional Features
Antimicrobial
Breathability
Easy to don and doff

Summary
Midi-Rox is a reversible wrap dress designed to empower independent one-handed dressing.

Design Goal
To create a garment that enables one-handed dressing and undressing from a seated position to minimize the restrictions of clothing fasteners and caregiver assistance.

Team Midi-Rox
Roxinne (Roxy) Gaussite: Client and subject-matter expert
Michael Tranquilli: Occupational therapist
Alyssa Wardrop: Fashion designer
Ray Luo: Engineer

Collaborator Acknowledgments
The Riverside Premier Rehabilitation & Healing Center, Open Style Lab™, Polartec®

Background
Individuals with upper extremity impairments are trained by occupational therapists in rehabilitative and/or compensatory strategies to resume functional activity. Rehabilitative strategies aim to restore lost functioning, while compensatory strategies aim to adapt procedures, devices, and environments to resume participation in ADLs. Although rehabilitative strategies to restore lost function were reinforced over the ten-week design process, the team's mission was to investigate adaptive strategies to improve performance in the activity of dressing.

Introduction

Roxinne (who goes by Roxy) is a retired psychologist who worked at the New York State Department of Health for more than twenty-five years. Since then, she has experienced a stroke—the leading cause of long-term disability in the United States, and high blood pressure (also known as hypertension)—the single most important risk factor for stroke, according to the American Heart Association. Roxy's stroke impaired the regions of her brain that are responsible for task-specific sequencing and memory skills. It also caused paralysis on the right side of her body, making it challenging to complete activities of daily living, such as dressing, bathing, and feeding without assistance from a caregiver, disrupting her personal and social lifestyle. She currently resides in the long-term care community at The Riverside Premier Rehabilitation & Healing Center in Manhattan, where she has expressed her desire for stylish clothing that enhances the quality of her life.

Design Process
1. Observation and Research
2. Goal Setting
3. Identifying Design Requirements
4. Prototype Iterations

Observation and Research

"Nothing raises my blood pressure more than having to wait for someone to help me get dressed."

Roxinne 'Roxy' Gaussite (pers. comm.)

The team members noticed Roxy having difficulty standing and keeping her balance due to paralysis in her right arm, leg, and trunk musculature. She requires 75 percent assistance from caregivers to stand, pivot toward, and be seated onto another surface (like moving to and from her bed and wheelchair). Unable to safely support her balance and simultaneously manipulate clothing, Roxy and her caregivers prefer she mostly dresses and undresses while seated in her wheelchair to reduce the risk of falling or injury to either party. The team's occupational therapist, Michael, previously conducted a data assessment. Data collected from the Upper/Lower Body Dressing components of the FIM (Functional Independence Measure) indicated that caregivers perform between 75–100 percent of the components required to fully clothe and undress Roxy to save time, promote safety, and conserve energy.

Goal Setting

Taking into consideration these performance and safety findings, the team committed to providing

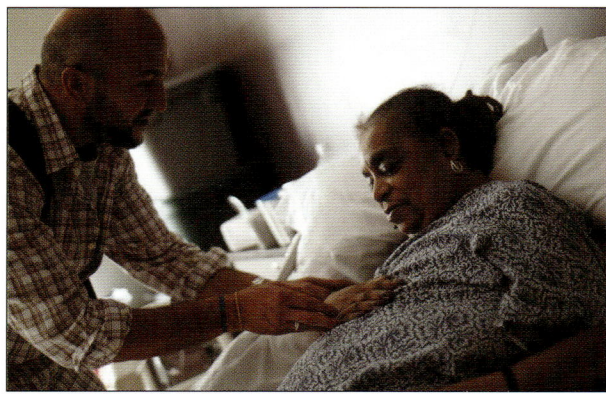

Figure 4.10a–b Photos of Michael Tranquilli and Alyssa Wardrop visiting Roxinne (Roxy) Gaussite at The Riverside Premier Rehabilitation & Healing Center in New York City. Courtesy of Kilian Son.

Figure 4.11 Michael Tranquilli, Alyssa Wardrop, and Ray Luo presenting their collaboration experience on adaptive fashion at Open Style Lab.
Courtesy of Open Style Lab.

Roxy with an accessible solution that reduced her dependence on caregiver assistance for dressing. They first analyzed the strengths and weaknesses of her existing wardrobe before proposing strategies that would allow one-handed dressing and undressing to be completed from a seated position.

Identifying Design Requirements
Roxy wasn't happy wearing what she called "plain and oversized clothing typically worn by disabled people" (Gaussite, pers. comm.) because it didn't reflect her personality or professionalism. So, she wanted the team to explore a mid-length dress with vibrant colors for attending social occasions with minimal or easy-to-use closures. Additionally, the team set out to create a garment with the following features:
1. Stretchy
2. Moisture-wicking
3. Fast-drying
4. Lightweight
5. Breathable
6. Warm
7. Easy to care for

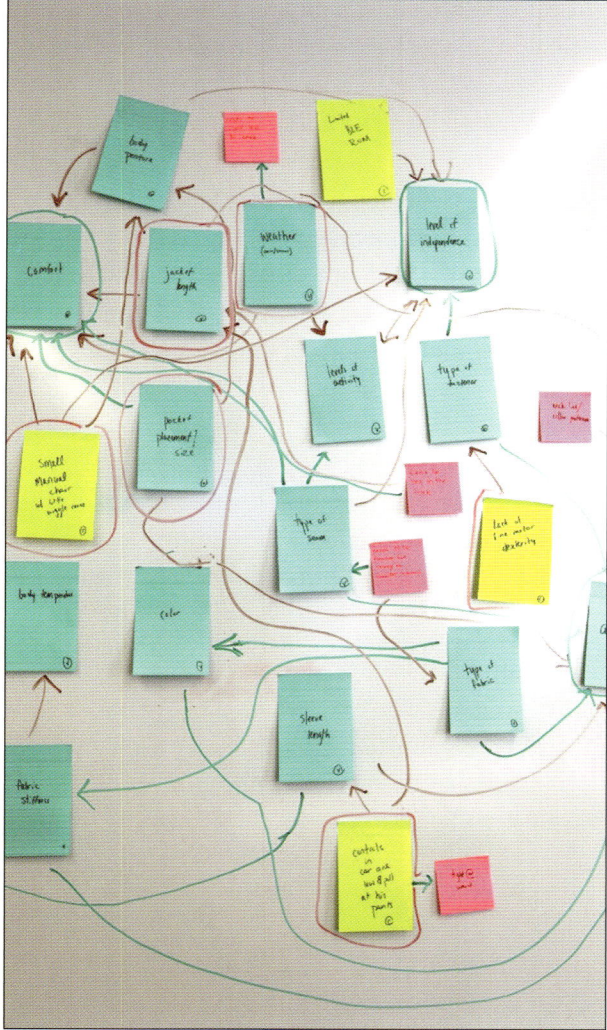

Figure 4.12 Design requirements identified by Team Midi-Rox. Courtesy of Michael Tranquilli.

Prototype Iteration 1
The team's first prototype was based on using the Over-the-Shoulder method of compensatory upper body dressing to eliminate the need for standing and the risk of falling or injury. This entails using the non-paralyzed arm to guide a jacket sleeve over the paralyzed arm and shoulder first. Holding the neck collar, the jacket is then guided around the backside before placing the non-paralyzed arm through the armhole and sleeve. After seated and standing body

Case Studies, Stories, and Interviews

measurements were taken, the team created a mid-length, jacket-style prototype in muslin—a plain-woven cotton fabric—to initiate testing with Roxy. They envisioned the dressing process beginning with caregivers placing the dress within the wheelchair—neck collar draped over the backrest and the sleeves draped over the armrests of her wheelchair. With the outside of the dress facing down, Roxy could sit atop her dress within the wheelchair and begin one-handed dressing. She only needed to guide her arms through the sleeves before overlapping the front panels to complete dressing. In this manner, her caregivers would ideally be able to set out the dress in advance. Positioning the dress behind and underneath Roxy would also eliminate the need for assistance maneuvering the dress around her backside, effectively removing one step of the Over-the-Shoulder method.

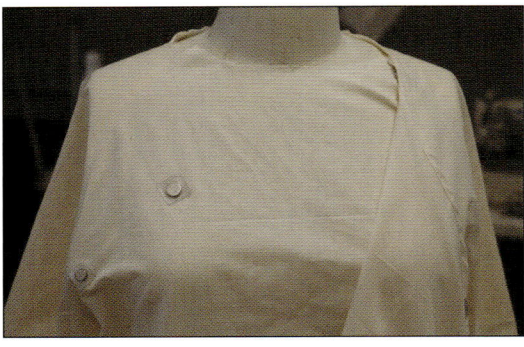

Figure 4.13a–b Muslin prototype with magnet placements to test for best opening spots on the garment. Courtesy of Alyssa Wardrop.

Summary

Two methods for dressing: around the back (coat or jacket) and over the head (T-shirt or sweatshirt). Roxy has a lot of hair to get through the neckline and wanted to reduce the possibility of messing it up when donning and doffing the garment. The team also enlarged the arm sleeves to decrease the level of accuracy needed to accommodate a paralyzed arm.

Features
- Enlarged neckline and armholes
- A vent for temperature control and wheelchair alignment
- A magnetic flap that lets the garment drape over the wheelchair for easy moving
- Magnetic closures for easy caregiver assistance

Prototype Iteration 2

After constructing and testing their initial prototype, along with video analysis, it was revealed that Roxy had improved performance using the Overhead method compared to the Over-the-Shoulder method of dressing her upper body. Specifically, reduced time, effort, and assistance were recorded. Over numerous trials, it became evident that donning a shirt by first placing the neck collar overhead was more intuitive for Roxy to initiate rather than dressing her impaired side first. She also displayed significant ease when locating the neck collar of a shirt compared to the armhole in the sleeve. Placing the garment overhead first created a draping of the fabric that naturally expanded the armhole. This sequence provided a static target location that required less effort to approximate and guide her arm through the sleeve.

Before dressing—as the dress was being placed within the wheelchair, the magnetic closures kept attaching themselves to the metallic frame of Roxy's wheelchair. This obstacle imposed more time and effort to detach the magnets and correctly reposition the dress, often requiring numerous attempts. Regardless of these efforts, after sitting atop her dress, the magnets frequently reattached to other metallic locations on the wheelchair. Attempting to locate and detach the magnets as a precursor to dressing became discouraging for all parties involved. Despite the simplicity and success, Roxy demonstrated how using this magnetic closure design to complete the dressing process still required assistance from caregivers—requiring help with the preliminary steps of detaching magnets, positioning armholes, and donning sleeves over the arms and shoulders.

Summary
Magnets stuck to the wheelchair, adding weight to the garment and making it difficult for Roxy to determine if she was wrapping the dress for a loose or fitted design. The back "flap" for ventilation in the garment design was unreachable for Roxy to close, requiring more time waiting for assistance and raising her blood pressure—although providing an easy caregiver setup. The team concluded that the neck and sleeve holes were appropriately sized to reduce the difficulty of getting the garment over the hands, arms, and shoulders.

Features
- Enlarged neckline and armholes
- A vent for temperature control and wheelchair alignment
- Magnetic closures for easy caregiver assistance

Prototype Iteration 3
Convinced that modifying those designs and procedures would accommodate shortcomings, another prototype was developed. The team couldn't overlook the safety risks for both Roxy and her caregivers when standing to dress. The team, therefore, chose to maintain the

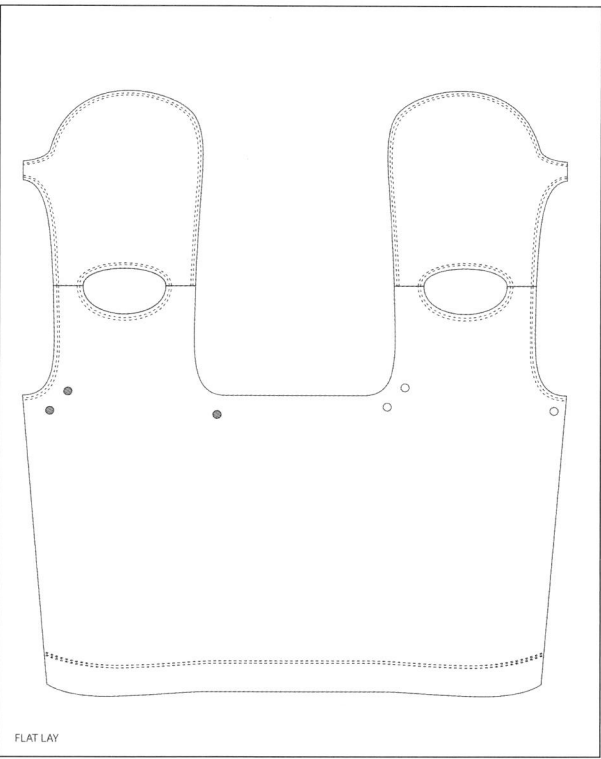

Figure 4.14 Pattern sketch by Alyssa of seamless wrap design with magnets. Courtesy of Alyssa Wardrop.

fall-prevention strategy of dressing from a seated position. Based on the Over-the-Shoulder method, caregivers would still set up the dress for Roxy to sit atop. However, to allow for convenient placement within the wheelchair, all magnetic closures were removed. Although Roxy was successful in manipulating the magnets to fasten and unfasten her dress, she continued having trouble using the Over-the-Shoulder method of dressing. In contrast, when using the Overhead method, Roxy was able to easily dress her upper body by first placing the neck collar overhead (like a sweatshirt) before guiding her arms through the sleeves. Based on these insights, the team combined elements of both the Over-the-Shoulder and Overhead methods, creating a new clothing pattern in a jersey-knit fabric—the Midi-Rox (see fig. 4.15)—based on a

Figure 4.15 Blue-colored prototype design utilizing existing garments. Courtesy of Alyssa Wardrop.

one-handed compensatory dressing method. The design was specifically designed to enhance Roxy's functional abilities by accommodating her intuitive dressing behavior.

Features
- Enlarged neckline and armholes
- A vent for temperature control and wheelchair alignment
- No closures for easier laundering and to avoid sticking to the wheelchair
- Layers to provide coverage and a feminine look

Material Studies

Because Roxy is seated throughout most of the day, choosing a breathable material with superior moisture-wicking qualities was essential for personal comfort and hygiene. It was equally important to use material with easy care instructions to minimize the time and cost of cleaning at the rehab center. After reviewing numerous material characteristics, the team chose the Polartec® Power Grid™ double-sided knit jersey (see fig. 4.16). Inclusive of the aforementioned elements, this material has a slight mélange weave on one side and grid texture on the opposite, which provides a variation of style for different occasions.

Materials
- Polartec® Power Grid™ double-sided knit jersey
- Magnetics enclosed in plastic

Figure 4.16 Double-sided Polartec® knit material in lime green and forest green colors. Courtesy of Alyssa Wardrop.

Final Design Outcome

To help Roxy identify and correct her postural imbalance when sitting and standing, the team investigated technological solutions to facilitate awareness and self-corrective behavior. What resulted was the Midi-Rox, a dress (1) crafted

Figure 4.17 Roxinne and Michael presenting the team's collaborative work at Parsons School of Design during the 2018 Open Style Lab summer program. Photography by Kilian Son. Courtesy of Open Style Lab.

according to a technical pattern and Roxy's body measurements to achieve a custom-tailored fit and (2) that combined elements of both the Over-the-Shoulder and Overhead methods (The American Occupational Therapy Association 2020). The Midi-Rox pattern is atypically fashioned with two neck collars on opposing sides, eliminating all closures on the dress. Roxy wears the garment by using a one-handed dressing method to grasp the neck collar placed overhead before withdrawing her left arm from the sleeve.

References

The American Occupational Therapy Association. 2020. "Occupational Therapy Practice Framework: Domain and Process Fourth Edition." *American Journal of Occupational Therapy* 74 (2): 1–87. doi:10.5014/ajot.2020.74S2001 (accessed August 14, 2023).

The Little Black Bag
Colleen Roche, Yiyun Guo, Kalyani Tupkary, Yue Zhang

Figure 4.18 Colleen Roche, Yiyun Guo, Kalyani Tupkary, and Yue Zhang with the final design of The Little Black Bag. Photography by Kilian Son.
Courtesy of Open Style Lab.

> **Key Functional Features**
> Portability
> Water wicking
> Easier reach

Summary
The Little Black Bag features a comfortable design that can be worn as a wheelchair accessory or crossbody bag.

Design Goal
To create a comfortable and versatile wheelchair accessory that accommodates the user's style and range of motion.

Team Little Black Bag
Colleen Roche: Client, subject-matter expert, and disability rights activist
Yiyun Guo: Fashion designer
Yue Zhang: Design technologist
Kalyani Tupkary: Design technologist

Collaborator Acknowledgments
Parsons School of Design, Open Style Lab™, Polartec®

Background
"Most power chair accessories are never stylish and almost always look like medical devices."
Colleen Roche (pers. comm.)

This collaborative project offers insights into the boundaries between assistive products and mainstream designs. It's often forgotten that people who use assistive devices, like wheelchairs and canes, also care about functional products that reflect their personal styles. PWD don't all desire invisibility through unobtrusive design and, in fact, want to be seen.

Wheelchair products have a stigma when their design and aesthetics aren't considered. The team prioritized discussions about ableism and the **medical model** of disability throughout the design process. Their goal was to create a rich and enabling collaboration to encourage a sense of belonging for each member. They focused on developing a strong relationship, conversing to better understand disability stigma, and celebrating disability creativity while designing a power chair accessory.

Introduction
Colleen is an activist and health educator. She's trained thousands of people on emergency preparedness, domestic violence, abuse, public health, and disability. She works from home and travels throughout New Jersey to deliver her training sessions. An avid photographer and

member of a Manhattan-based integrated dance company, Colleen is always on the move, commuting to several locations within the city.

Colleen was born with cerebral palsy and uses a powered wheelchair that she's traveled thousands of miles in over the course of four years. She identifies as non-ambulatory, and her chair a secondary skeleton—her most visible, valuable accessory and the place she spends most of her time. Because dressing requires an inordinate amount of time and energy, the outfit she dons in the morning is nearly always what she wears until the evening. "The extent to which our disabilities impact us from day-to-day really can vary significantly. This, of course, also affects what we choose to wear and how we use products," says Colleen (Roche, pers. comm.). Pain, range of motion, and edema, particularly in her lower extremities, can vary significantly for Colleen throughout the day. Therefore, she typically favors clothes that are comfortable, simple, and can transition from work to social events.

Design Process
1. Observation, Research, and Discussions About Ableism
2. Goal Setting
3. Identifying Design Requirements and Challenges
4. Prototype Iterations

Observation, Research, and Discussions About Ableism
The team first looked at the range of items Colleen typically carried with her in her wheelchair. Contents inside the narrow side bag on the armrest (usually overstuffed and bulging at the seams) included:
- Phone
- Wallet
- Cards
- Pens

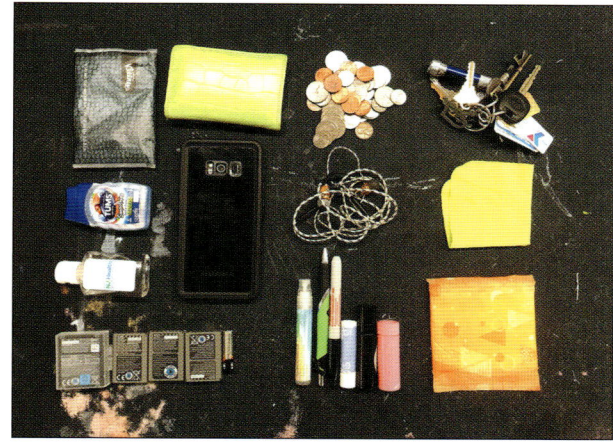

Figure 4.19 Items found in Colleen's bag, such as phone, keys, and battery packs.
Courtesy of Colleen Roche.

- Napkins
- ChapStick
- Mirror
- Keys

Inside her backpack (which required twisting, grabbing, and heavy lifting):
- Laptop
- Camera
- 1–2 water bottles

They also took note of her clothing style—mostly stretchy materials in monochromatic neutral tones with a few pops of color and repetitive patterns, and a few products she had already modified and adapted. For example, she designed her own rain cover to protect the electronic steering wheel of her chair by using a plastic umbrella bag.

While not particularly stylish, it successfully solves the problem. In addition, Colleen has discovered ways to use (1) home furniture to elevate her electric cables and other items off the floor for easier reach, (2) silicone rubber grips for doors to make them easier to open (The American Occupational Therapy Association 2020), and (3) magnet clasps to easily put on delicate jewelry.

Case Studies, Stories, and Interviews

Figure 4.20 Adaptations found in Colleen's home. Courtesy of Colleen Roche.

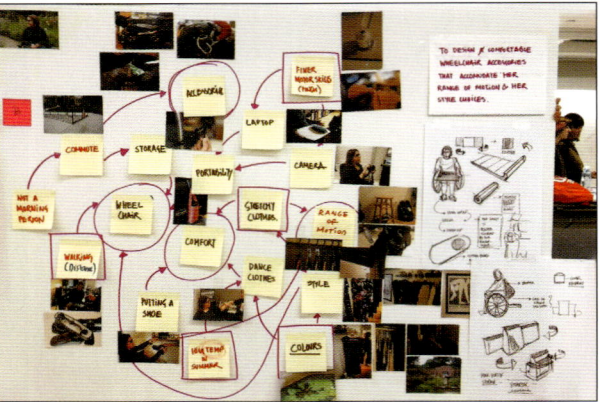

Figure 4.21 Sketch of the team's design requirements board. Courtesy of Yiyun Guo.

Goal Setting

The team then began exploring a variety of accessory designs that were aesthetically pleasing and able to provide ample storage space. They set out to create a final product that was compact, comfortable, waterproof, and sturdy.

Identifying Design Requirements and Challenges

There proved to be several challenges with Colleen's current carrying system. For instance, the heavy backpack on her chair slowed down the speed of her wheelchair. The team's solution focused on portability with size restrictions, encouraging Colleen to limit the number of items she traveled with.

Prototype Iteration 1

"I'd like to find something that lets me have more control over my personal items."

                                Colleen Roche (pers. comm.)

The first prototype wasn't a bag but a foldable fabric desk created by Yiyun. Because Colleen's wheelchair positions her lower than most tables, she tends to work from and eat off her lap. The team wanted to ease this frustration by designing a lightweight structure that could easily be rolled up and stored away. What resulted was a modular desk bag—a backpack with a fold-out table made from fabric. It remained hooked onto the wheelchair when not in use and featured a side bag that could flip onto her lap for easy access. However, it didn't accommodate Colleen's range of motion, as it was hard for her to reach and take out.

Figure 4.22 Sketch of foldable and collapsible desk that was part of a modular bag design idea sketched by Yiyun. Courtesy of Yiyun Guo.

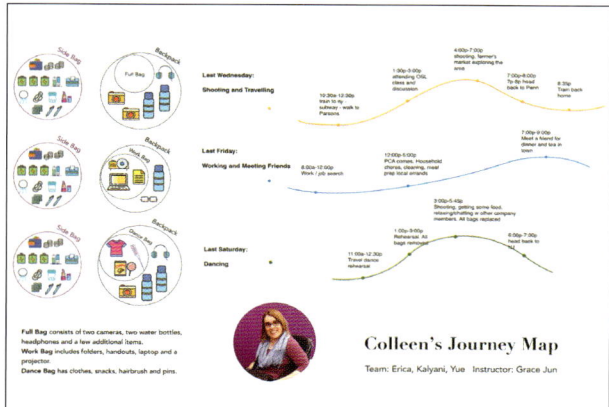

Figure 4.23 A journey map sketch of Colleen's commute, drawn by Yue. Courtesy of Yiyun Guo.

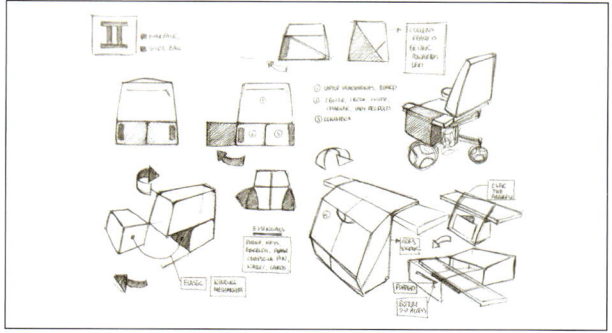

Figure 4.24 Sketch of bag design idea sketched by Yiyun. Courtesy of Yiyun Guo.

Next, Yue created a journey map to help the team better understand the issues surrounding weight, size, and portability. This helped them prioritize their design challenges and realize that they should make a side bag for her essentials that could slip onto Colleen's power chair arm.

Prototype Iteration 2
The design focus of the second prototype stemmed from the fact that Colleen's bag was always removed and handed off to someone else every time she needed to be lifted out of her power chair (e.g., onto a plane seat or into an inaccessible bathroom stall). This meant that she couldn't always keep her personal items close. So, the team started on a crossbody bag design that featured buckles and a strap. Also convenient for the mainstream market, it could be worn seated, standing, or strapped under the armrest of a wheelchair (making it easier to navigate smaller spaces).

This design began as a box-like structure that opened from the side with a zipper that curved all the way around it. However, the team quickly realized it was hard for Colleen's hands to grasp its items sideways, and therefore made the following modifications:

Figure 4.25 Second bag iteration using a soft polyester fabric with a hold ring on the zipper. Courtesy of Yiyun Guo.

1. Adding extra outer pockets for specific items (e.g., phone).
2. Adding an extra flap with Velcro® on the top of the bag to secure it onto the armrest.
3. Replacing the ring holes with larger ones to create an easier grip when unzipping the bag.

Prototype Iteration 3
At this stage, the zipper was still difficult to pull around the bag's sharp, angular corner. So, the team replaced it with a straight zipper at the top

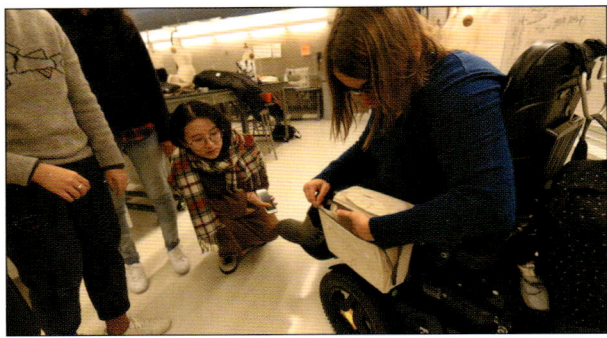

Figure 4.26 Colleen testing the muslin bag prototype for accessible reach. Courtesy of Grace Jun.

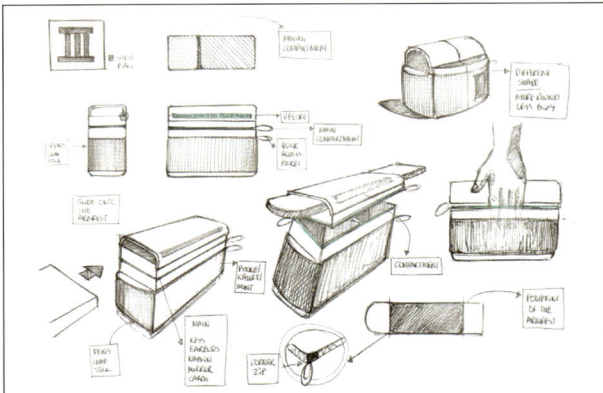

Figure 4.27a–b Cutout bag patterns backed on canvas for a structured look with sketches of the bag's envelope style by Yiyun. Courtesy of Yiyun Guo.

of the bag and redesigned the opening with an accordion fold like an expandable envelope. Other small additions included pockets on the inside and double Velcro® for reinforcement.

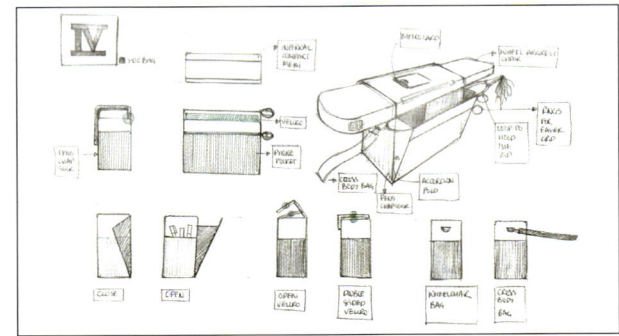

Figure 4.28 Final sketch of the bag iteration design with accordion fold-out. Courtesy of Yiyun Guo.

Prototype Iteration 4

One the muslin prototype was created and solidified with functional features that accommodated Colleen, the team turned their attention toward the bag's aesthetics. Based on her input, wardrobe, and style, they chose a black monochromatic fabric with marble accents. The lightweight, high-performance material was waterproof with a canvas backing to help ensure a strong structure that would withstand the rain, and the light-colored interior lining made it easy to identify the items inside.

Materials
- Cotton
- Muslin
- Large circular black zippers
- Polartec® NeoShell®, a water-resistive fabric
- Custom-dyed fabric with marble-like patterns

Final Design Outcome

"This bag, which is carefully and efficiently designed for a wheelchair user, is a product that anyone would want with very little modification."

Colleen Roche (pers. comm.)

The rectangular side bag is strapped onto Colleen's power chair armrest with heavy-duty Velcro® but can be detached and worn as a crossbody bag. Its

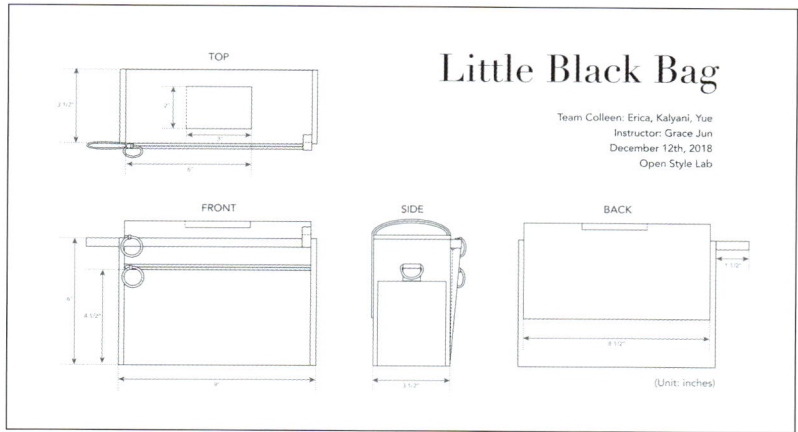

Figure 4.29 Technical drawing of the bag design. Courtesy of Yiyun Guo.

shape offers sufficient storage space while still being compact, which enables her large assistive device to maneuver small, narrow spaces (e.g., doorways and corridors). The interior is divided into two compartments for larger items, and the exterior features an open pocket for ChapStick and pens, a zipped pocket for her phone, and a small slot at the top for her credit cards.

Since the project concluded, the team has continued to check in on Colleen and her bag to make minor repairs and discuss maintenance considerations. "Over the last three years, Colleen and I have had intermittent correspondence. Each time, she lets me know how the bag is doing. Knowing that Colleen still finds [the bag] useful has been rewarding," says Kalyani (Tupkary, pers. comm.). Overall, the design process and final

Figure 4.30 Crossbody function of the bag. Courtesy of Yiyun Guo.

product demonstrate that considering aesthetics and personal style helps increase product compliance and ensure longer use.

References

The American Occupational Therapy Association. 2020. "Occupational Therapy Practice Framework: Domain and Process Fourth Edition." *American Journal of Occupational Therapy* 74 (2): 1–87. doi:10.5014/ajot.2020.74S2001 (accessed August 14, 2023).

Figure 4.31a–b Side view of the bag with the Velcro® fastener to latch onto the arm of Colleen's wheelchair. Front view of the bag with two enclosures supported by ring hooks attached to the zippers. Courtesy of Yiyun Guo.

Interview
Understanding the Use of Materials— Angela Domsitz Jabara

Figure 4.32 Angela Domsitz Jabara, Education Director at the AATCC. Courtesy of Angela Domsitz Jabara.

With a background in apparel and textile design, Angela Domsitz Jabara is the Education Director at the AATCC (American Association of Textile Chemists and Colorists). She had prior experience working as a Project and Education Manager at The Woolmark Company and Fabric Sourcing at PVH Corp.

Grace
Could you please tell me a little bit about yourself and how you got involved in the work you are doing today?

Angela
I am currently the Education Director at the AATCC and live in North Carolina. My mom introduced me to textiles when she started sewing dresses. I've worked in fashion textiles for over fifteen years now, across different roles within the supply chain. My job right now is working with a lot of people in the chemistry and color side of the business, which includes finishes, dyeing, and the development of new materials. It's been interesting to test textiles. Prior to my current role, I was deeply involved with Australian wool growers. I had a great chance to learn about the fiber side of the supply chain role, which is the start of most supply chains. I then worked for brands like PVH and Elie Tahari on the fabric development and production side of the business. I studied apparel and textile design at Michigan State. I just love working with textiles. I think it's an amazing opportunity to meet new people, to travel, and to learn all the time because textiles are always changing. There's always something new, and it's so exciting to get to experience this. Once you get into it, it's like everything.

Grace
Materials have a profound impact on the adaptive fashion design process. Fabrics and other materials worn can either constrain or expand a wearer's movement. Could you expand on material testing and what designers could be more aware of when making?

Angela
Testing is always evolving because there's always a new thing we need to know or a new aspect of a material that needs to be uncovered. Material testing has different reasons for developing, such as it being tested for marketing purposes. What are the claims that we can share with consumers? How can it be, in a scientific way, repeatable? You want to be able to test for the actual product you're selling. Is it made of what you're claiming it is? Testing for claims and testing for the type of product you're selling are some of the main reasons. And there is also performance testing. You want to know what this material can do. How does it meet the standards that I think it should meet for my consumer? Factors like tear strength, abrasion, wrinkling, or appearance testing are included. Are these things maybe not considered by most people who buy products, but they're heavily considered by people? So, in order to meet those expectations, companies have to be able to validate products that are sold regularly and have credible data. I work with folks, committees, and volunteers from all over the industry. They come together to talk about the challenges in textiles, such as developing scientific test methods for textiles. And they don't get paid for this. This is something that anyone can participate in, and you don't have to be a scientist. You just must be interested in it.

Grace
You mentioned consumer expectations. In the case of consumers living with disabilities, how do you think testing performance materials can be beneficial?

Angela
Testing and performance go hand in hand. Brands and companies are focused on testing because performance materials imply there's some inherent quality that a material will perform to a certain standard. It's important that you understand how a material works as a product developer or designer. Consumers want to trust people selling something and know that it is what it claims to be. For consumers living with disabilities, a material can provide protection from the environment and offer benefits like comfort, warmth, or cooling. Some fabrics have textures that support the body against abrasion. Abrasion-resistant materials or materials that have high tear strength reinforced in certain areas, such as parts of heavy use areas of a product, are very useful. Testing materials also provides insight into product care and laundering.

Grace
Regarding laundering, could you expand a little bit more about wrinkle-free material?

Angela
To preface this question, materials can be constructed in many ways. They're constructed from fiber, yarn, and the actual design of the material, and then finish also impacts the way it's constructed. At any point in that process, you can impart functionality, which is the cool thing about textiles. Something like wrinkle-free will be imparted in different aspects. For example, wool is naturally wrinkle-resistant and can be shed easily through exposure to moisture vapor because the fiber absorbs moisture into it, and it releases any creases. If a fiber is not naturally wrinkle-resistant, like cotton, you would look at that functionality by applying a finish to it. Typically, when the fabric is complete, you would apply certain types of finishes to it, which allow it to resist wrinkling.

I worked in men's shirts for a long time, and wrinkles are a big part of that business. Cotton, a typical fiber choice for shirting, has finishes applied to it.

Grace
I've experienced stretch fabrics to be extremely useful not just for myself but for many people. Could you explain a fabric's stretch properties and how it could help people get dressed more easily?

Angela
Stretch can be imparted at different moments within the textile's life. There are fabrics that are inherently stretchy, and then there are fibers that have stretch designed into them. For example, synthetic fibers can be manufactured in a way that incorporates stretch to them. Wool is a material that has natural stretch. So, depending on the level needed for the user, designers have many fiber choices. This is dependent on the amount of stretch and direction of stretch needed, as seen in two-way stretch from four-way stretch materials. You must think about, which direction do people need that stretch?

I think a lot of people consider knitwear when we think about stretch because knitwear inherently comes with stretch properties. Stretch can be achieved in the yarn or the fiber stage. Knit material is usually the best bet in a fabric stage but, depending on the density of that material, it might not have as much stretch if the yarns have an open structure. Many times I experience meeting people who think stretch means a material is less durable and tears easily. For example, yarn fibers can get snagged more easily and quickly, unraveling the material. Then there are some types of yarns that degrade over time, especially when exposed to heat, like spandex can change color to yellow as it degrades over time.

Grace
When Open Style Lab™ was designing a jacket with Jim Wice, Director of Disability Services at Wellesley College, his interdisciplinary team wanted something breathable. As you know, Jim, who has a spinal cord injury, was extremely mindful of temperature regulation for his body. The team was provided with a green wool material that was thin and light in weight, perfect for a suit jacket from the Woolmark Company. How does this fabric work, and what other benefits does it have?

Angela
That type of wool fabric is a woven and has natural stretch. Because of how it's manufactured, it can stretch in different ways. Like socks elongate to the point at which they do not return to their original stretch. A natural fiber will return to its original shape. It's great for people who maybe want to spend a little bit more on a piece, but it is so functional to use for years and years. It will maintain its appearance, which is important not only for the look of the garment but also for its use.

It's important to consider the shape of the person wearing it, their needs, and mobility. As the person moves, this fabric is great because it provides natural resistance to liquid, water, and moisture. This particular fabric the team used is also used for swimwear, in swim trunks. It's also used in rainwear. It's not going to protect you if you go out into an extreme downpour, but it will keep you somewhat dry. This fabric can be worn as an everyday item. It would help keep someone protected from the elements. In addition to that moisture management, it absorbs vapor, so humidity is released naturally, making it very breathable. It is also wrinkle-resistant and biodegradable. Lots of benefits.

Figure 4.33 Merino Perform WP from Nanshan, provided by the Woolmark Company. Courtesy of Open Style Lab.

Figure 4.34 Jim Wice, Director of Disability Services at Wellesley College, wearing SUITable, a thermally adaptive sport coat featuring easy-to-access front ventilation flaps, as well as magnetic closures with hidden pockets for adjustability by individuals with limited dexterity. Designed by the 2016 fellows at Open Style Lab. Courtesy of Open Style Lab.

Grace

A frequent question I receive from students is, how can I get performance fabrics? How do I get to the facilities that can help me test these materials when trying to design something with them? Do you have any thoughts and tips for students or designers working with complex and scientific materials?

Angela

That's a great question, especially for designers who are new to the industry. The access to certain fabrics is sometimes only available to those who have large businesses. Many of these large businesses might already have their own facilities that produce these materials. Performance materials are a challenge because they require a lot of thought to develop them. For example, we discussed earlier about the different layers of textiles where finishes that have performance properties can be at those different levels. Not everyone always has access to do that—place functional properties in different layers of a textile.

There are some things that are more accessible. I've worked with designers who created a wax wool jacket. Wax as a treatment on cotton is not uncommon, but these designers wanted to do it on wool because that fiber has a natural water resistance, and adding wax to it is a natural way to enhance performance. Wax is basically a fat, you know, so just like any other fat or oil-based product, like a nylon or polyester, it can impart some of those protective barriers against the elements. Not all these finishes in fabrics are bad for the environment. Biometrics or design thinking about biology could inspire some people to think differently about textiles and how maybe some of the things that we already have around us in nature are potentially used.

Case Studies, Stories, and Interviews

Grace
Not all students have access or economic resources. What resources would you recommend to students?

Angela
That is true. We've worked with students, and some don't have access to a shop. For example, local Joanne's is three hours away. I think sewing has become less common in the United States. People aren't buying fabric as much as people aren't buying trims. They just buy a garment in a store, and the supply chain in the United States is very different from a global point of view.

    I recommend developing and building your network. Finding people on LinkedIn makes it so easy to look around and ask people to see what's out there. At AATCC, we have volunteer groups where students can join the student chapter. While this organization has a high focus on chemistry and color, and testing, it's still a great way to build a network and access resources as a student.

Grace
Which schools would you recommend to students wanting to explore performance materials and textiles more deeply?

Angela
It can vary from the design of textiles to chemistry and science. North Carolina State (the best textile school in the United States), Cornell, and Oregon State have great programs for their performance apparel and products. The Fashion Institute of Technology has great textile studies with many resources, and design programs at Parsons School of Design are also great ways that students can explore. Building your network through social media or going to events in the textile industry, like trade shows, are also alternative ways. Students have discounts and can have access to many conferences. One of the experiences that got me inspired to be in textiles was a visit to the Cooper Hewitt Museum during college. They had a design and textile exhibition, where I saw how the functionality of textiles can have different applications. Museums are not only a data resource but rich with new inspiration.

Grace
It seems like there are many variables in textile making, but this shouldn't prevent creatives from iterative design. How often do you think creatives need to consider natural resources and labor-intensive processes when using a material?

Angela
I encourage everyone to make a textile at least once and understand what that process is. Going to a textile facility or manufacturer is one of the best experiences. I wish designers would do this more often because not understanding the limitations of a design is usually caused by many material factors. If you understand those material variables, you have a better understanding of how to make something.

Trans-Skirt
Rachel Handler, Eraince Wang, Helena Avraham, Juliette Stephanie Van Haren, and Nuomeng Zhang

Figure 4.35 Team Trans-Skirt from left to right: Juliette Stephanie Van Haren, Helena Avraham, Rachel Handler, Eraince Wang, and Nuomeng Zhang.
Courtesy of Open Style Lab.

Key Functional Features
Anti-microbial
Moisture repellent
Discreetness
Comfort

Summary
Trans-Skirt is a reversible, asymmetrical garment that helps make moving around the city comfortable for people with a prosthetic leg. It can also be reconfigured and used in formal settings and events.

Design Goal
To create a piece of clothing that is easily modified to hide or reveal different parts of the body.

Team Trans-Skirt
Rachel Handler: Client and subject-matter expert
Eraince Wang: Design technologist
Helena Avraham: Fashion designer
Juliette Stephanie Van Haren: Product designer
Nuomeng Zhang: Design technologist

Collaborator Acknowledgments
Parsons School of Design, Open Style Lab™, American Woolen

Background
One of the benefits that Congress extended to US veterans under the G.I. Bill in 1944 was access to prosthetics. However, unlike the cars and houses that provided some sense of normalcy, designed parts of the body were a particularly complex issue back then. Debates surrounding prosthetics were often on the spectrum of supportive rehabilitation for social integration versus excessive entitlement that may hinder a veteran's ability: "Arguments over limb design and distribution centered on questions of the relative responsibility of individuals and their technological tools" (Williamson 2019, 26). Unfortunately, this postwar cultural tension is still felt today by people living with disabilities.

Introduction
Rachel Handler is a filmmaker, actress, singer, and performer living in New York City who's won the AT&T Underrepresented Filmmaker Award for her short *Committed* and the Sundance Collab Monthly Challenge for her script *The A Doesn't Stand for Accessible*. Since joining the disabled community, she's found a passion for writing,

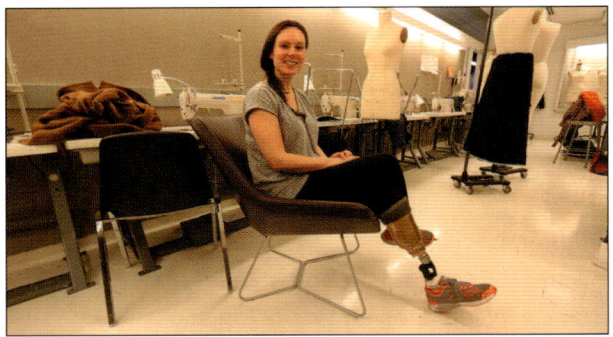

Figure 4.36 Rachel Handler wearing a gold prosthetic leg. Courtesy of Grace Jun.

producing, and directing while advocating for inclusion in every project she creates.

In her early twenties, Rachel lost her left leg in a car accident. She was able to salvage her knee, increasing her ability to maneuver her BKA (below-the-knee) prosthetic leg. However, she regularly experiences sweating around her amputated knee, resulting in balance problems. For these reasons, it's essential that she wears clothes that don't trap heat or hinder movement, and she prefers that they don't call unwanted attention to her prosthetic leg.

The Project

As a professional in the film and theater industry, Rachel often needs to quickly change clothes in small spaces. So, the team set out to construct a garment with her that was comfortable, allowed movement, and could easily be reconfigured to hide and reveal her prosthetic leg (depending on what she preferred that day).

Design Process

1. Observation and Research
2. Goal Setting
3. Identifying Design Requirements and Challenges
4. Journey Map
5. Prototype Planning and Development

Observation and Research

The team started by discussing disability, body, and design, quickly learning from Rachel that people with similar disabilities can have different needs. For example, while some individuals with prosthetics want to hide them, Rachel likes to show hers off—specifically without unwanted attention and reactions from strangers. This fact made it clear just how important it is to communicate with the client rather than assuming what they want or what they like. Giving her agency in this matter helped prevent "implications of normalizing appearance versus emphasizing social uniqueness" (Kaiser 1985). In a short amount of time, Rachel significantly helped the team better understand the reality of varying perspectives, concerns, and needs, which reinforced their need for her help with every step of the design process. Focusing solely on Rachel, they employed user-centered design principles

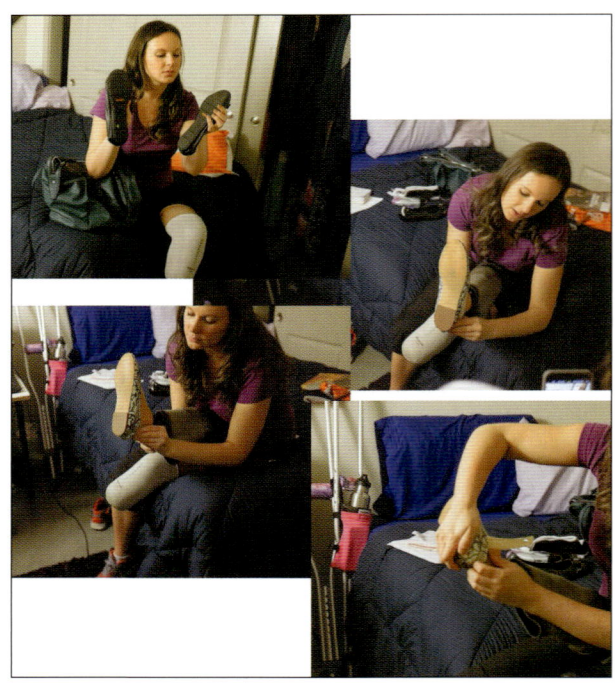

Figure 4.37 Integrating Rachel's expertise into the design process through observation of shoes and dressing. Courtesy of Eraince Wang.

Fashion, Disability, and Co-design

and methods that proceeded with behavioral observations, interviews, questionnaires, and usability tests.

Goal Setting

"I love maxi dresses when I don't want people to notice my leg."

Rachel Handler (pers. comm.)

By shadowing Rachel, the team learned about her lifestyle, commute, and wardrobe style—all largely influenced by her career, which consists of performing and meeting with casting directors. The most distinctive characteristic of her clothes is their color palette: pink, purple, and blue. Rachel also has a strong preference for leggings and bright dresses, both monochromatic and floral.

What she doesn't like are thick, heavy fabrics that make her knee hurt and/or sweat. When her prosthetic leg becomes too painful, she likes to take it off—however, regular pants, like jeans, make that hard to accomplish. The team's observations of Rachel and research on the relationship between materials and the body inspired them to explore a pant-skirt hybrid for her.

Identifying Design Requirements and Challenges

When commuting to auditions or performances, Rachel likes to wear yoga pants and sneakers but brings a change of semi-formal clothes, a pair of semi-formal shoes, and a shoehorn. She showed the team how she swaps outfits, which entails sitting down to take off her shoes and occasionally her prosthetic leg. The team immediately realized how time- and energy-consuming the process is and set out to address this pain point with three specific design requirements.

Design Requirements
- Able to be reformed and reshaped as needed
- Made with lightweight materials
- Matches Rachel's bright, bold, and confident sense of style

Journey Map

The team first created a visual journey map that reflected Rachel's typical routine, which illustrated the following insights:

1. She travels by public transportation wearing clothes that cover her prosthetic leg to avoid unwanted attention.
2. She carries with her a bag of clothes and shoes to change into.
3. She prefers to wear garments at auditions that show off her prosthetic leg because some roles are specifically written for actors and actresses with one.
4. She often changes her clothes in a restroom that's small and not accessible, making it even more difficult to doff her shoes, pants, and prosthetic leg.
5. She usually uses a chair for extra support during the process, which, on average, takes her about five minutes.

Figure 4.38 Colors identified to use for the design based on Rachel's wardrobe. Courtesy of Eraince Wang.

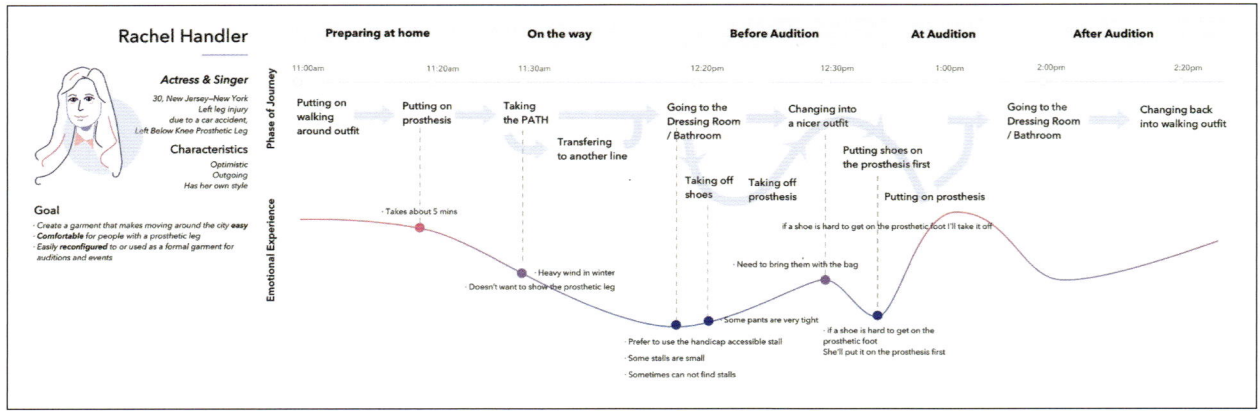

Figure 4.39 Rachel's journey mapped and designed by Eraince Wang and Nuomeng Zhang. Courtesy of Eraince Wang.

Prototype Iteration 1

The first garment was a pant design that only revealed one leg; however, it was hard for Rachel to put on and took even longer than her current donning and doffing process. So, the team decided to explore a skirt-shorts hybrid.

Prototype Iteration 2

The skort—a hybrid between a skirt and shorts—provides modesty and movement for wearers. It can be constructed as a pair of shorts with an overlapping fabric panel made to resemble a skirt or designed as a skirt with a hidden pair of integral shorts. In fashion, skorts are most often associated with bicycle riders or female tennis players.

The team realized its versatility could greatly benefit Rachel, so they iterated several configurations to better understand how intuitive it was to put on. While Rachel acted as the primary tester, the team also asked strangers in Washington Square Park to try on their skirt design (without any instructions) before giving them feedback. These public engagements provided invaluable insights that specifically addressed the following adjustments:

1. Adding straps to the pants with matching holes that inform the wearer how the garment should be assembled.
2. Attaching flaps onto a dress or concealed pants that let the wearer easily pull it on.

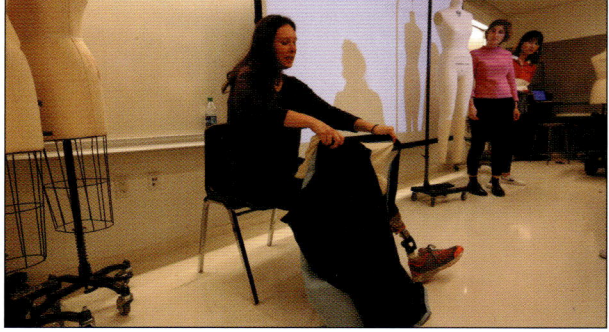

Figure 4.40 Rachel trying on the first prototype design. Courtesy of Grace Jun.

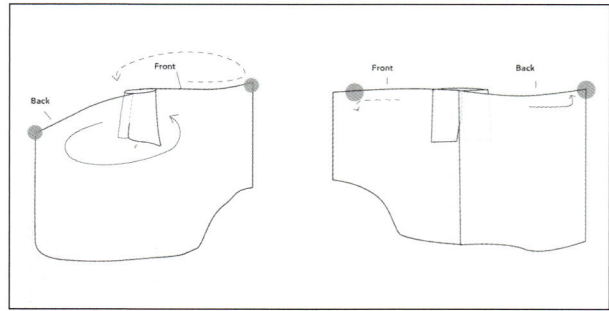

Figure 4.41 Transformative short design illustration, drawn by Helen. Courtesy of Helen Avraham.

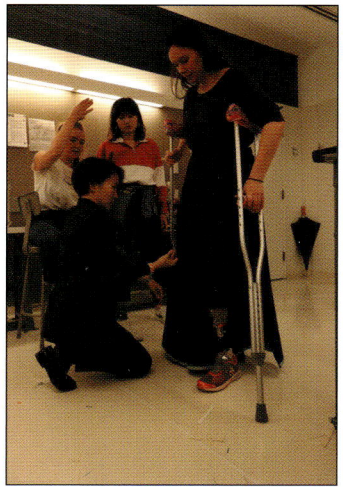
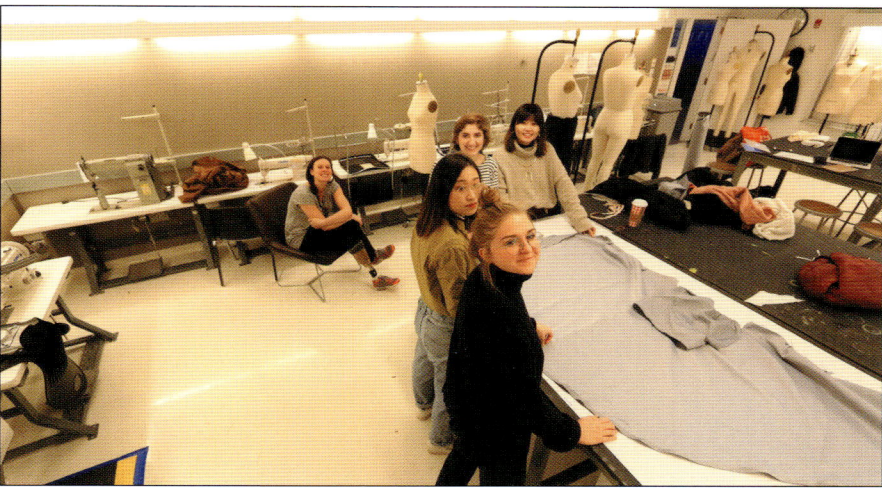

Figure 4.42 Rachel standing to test if the hem of the skirt would drag on the floor while walking.
Courtesy of Open Style Lab.

Figure 4.43 Photo of the Trans-Skirt team members before cutting into the final material. Courtesy of Grace Jun.

Prototype Iteration 3

After the team explored creating a silhouette with muslin prototyping, they began considering what fabric they would use. While smooth materials, like satin, made the donning and doffing process easier, Rachel needed something that was lightweight, breathable, and appropriate for both casual and semi-formal settings. Together, they all decided on denim for its aesthetics and durability before choosing a wool denim from American Woolen Company with properties that support regulating body temperature.

Final Design Outcome

Rachel's Trans-Skirt is a skirt-pants hybrid that uses the exact same cotton warp yarn as traditional denim but replaces some of the weft yarns with machine-washable wool. This unique fabric has all the benefits of a purely woolen garment—naturally flame-retardant, water-resistant, and insulated—and perfectly accommodates the needs of Rachel's body and the demands of her career. The final design is reversible, asymmetrical, and easily adjustable to either hide or show off her prosthetic leg.

Figure 4.44a–b American Woolen fabric that looked like denim.
Courtesy of Open Style Lab.

Case Studies, Stories, and Interviews

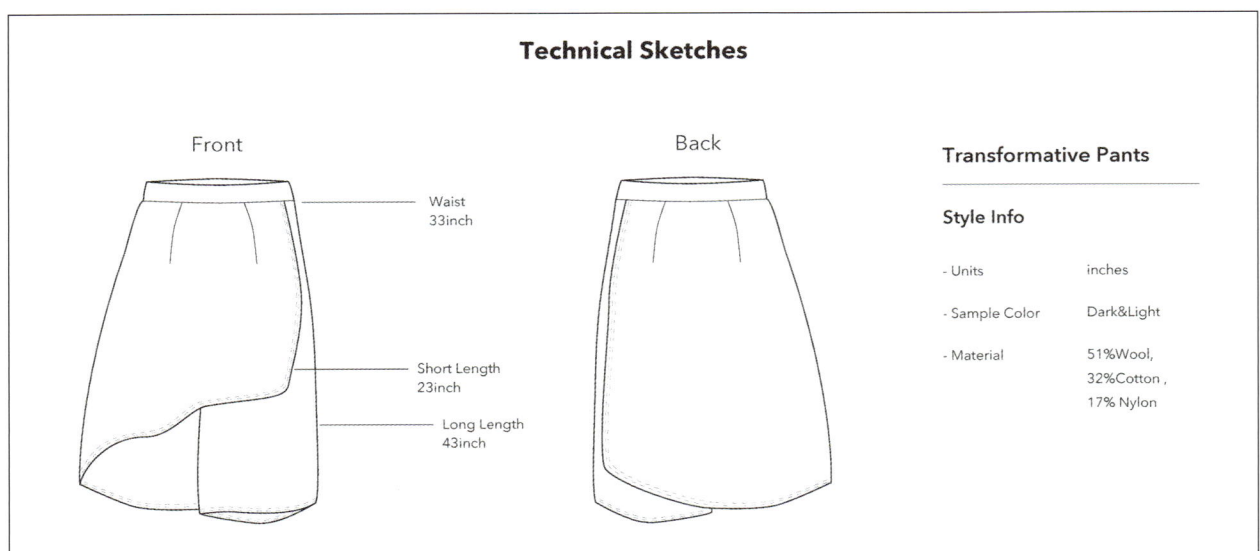

Figure 4.45 Final technical sketch of skort design. Courtesy of Helen Avraham.

Figure 4.46 a–c Rachel wearing Trans-Skirt final design. Courtesy of Open Style Lab.

References

Kaiser, Susan B., Carla M. Freeman, and Stacy B. Wingate. 1985. "Stigmata and Negotiated Outcomes: Management of Appearance by Persons with Physical Disabilities." *Deviant Behavior* 6 (2): 205–224, doi:10.1080/01639625.1985.9967670 (accessed August 14, 2023).

Williamson, Bess. 2019. *Accessible America: A History of Disability and Design*. New York: NYU Press.

qXgo
Quemuel Arroyo, Staci Chan, Kailu Guan, Chengcheng Zhao

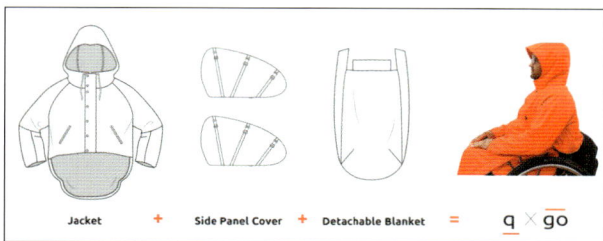

Figure 4.47 The design components of qXgo: a jacket, wheelchair panel covers, and a detachable blanket to cover the lap.
Courtesy of Chengcheng Zhao and Kailu Guan.

Figure 4.48 Team qXgo from left to right: Staci Chan, Kailu Guan, Quemuel Arroyo, and Chengcheng Zhao.
Courtesy of Open Style Lab.

Key Functional Features
Portability
Breathability
Durability
Water-repellent

Summary
qXgo is a two-piece raincoat that's breathable, portable, and waterproof for wheelchair users on the go.

Design Goal
To create a protective piece of outerwear with a holistic, systems-based approach.

Team qXgo
Quemuel Arroyo: Client, subject-matter expert
Staci Chan: Occupational therapist
Kailu Guan: Fashion designer
Chengcheng Zhao: Designer and engineer

Collaborator Acknowledgments
Open Style Lab™, Polartec®, Professor Min Zhu at FIT (Fashion Institute of Technology)

Background
Accessible public transport benefits all people in society. Found in many major US cities, ramps and curb cuts are some of the many examples of accessible designs created for and by people living with a disability. Despite these designs, a lack of accessible transportation is often cited as a barrier to employment for PWD in the United States (Bezyak, Sabella, and Gattis 2017; Wong 2020). Multiple factors contribute to disabled individuals avoiding public transportation, such as the spatial inaccessibility of subway entrances, long commute times, and exposure to unfavorable weather conditions (Wong 2018). Little is known about the commuting pattern of PWD and rarely is fashion proposed as a solution. Decisions about what to wear before leaving the house are crucial. Protecting a commuter with a disability throughout the public transit journey includes a combination of design factors, including materiality, body mobility, and the context of public spaces. Entering a subway entrance that has no direct elevator access during rainfall can be detrimental. Not only is

time wasted in finding accessible public transit entrances and exits, but weather conditions present a particular problem. Exposure to cold temperatures and rain can cause hypothermia in people with SCI/D (spinal cord injury or damage). This can potentially lead to pain and slowing of motor skills. Adaptive outerwear provides one solution for commuters, in particular for wheelchair users and people with SCI/D.

Introduction
Quemuel, nicknamed Q, is an experienced mobility expert living in New York. In 2007, during an outdoor sporting accident, he sustained a T9–T10 incomplete spinal cord injury, which typically impacts the motor, sensory, and/or autonomic function based on the location of damage (Christopher & Dana Reeve Foundation 2017). Despite this, Q has a highly productive lifestyle: He's a government official who oversees all the Department of Transportation's projects related to accessibility; a dancer with the Heidi Latsky Dance company; a community volunteer; and someone who loves swimming, rock climbing, and quality time with his family and friends. He independently completes his basic ADLs, such as bladder and bowel management, and occasionally uses adaptive equipment (a reacher) to grab things. Q also manages IADLs (instrumental activities of daily living), such as cooking and using the subway and Uber, at a modified independent level.

Goal Setting
"I'm always up for challenges, but the one thing that slows me down is the rain."
                              Quemuel 'Q' Arroyo (pers. comm.)

Q enjoys leading an exhilarating life full of routine and novel experiences. Yet, one aspect that challenges him is the rain. To protect himself from rain or snow, Q places a golf umbrella diagonally into his shirt while laterally tilting his head to stabilize it. While Q's solution works in open spaces, it becomes more challenging with other people. In crowded spaces, Q must continuously stop to self-propel and reposition his umbrella during his commute.

Design Process
1. Research
2. Observation and Interview
3. Goal Setting
4. Identifying Design Requirements and Challenges
5. Design Iterations and Prototypes

Research
"There is little known about adaptive clothing by the medical community or the needs of adaptive clothing by individuals with SCI" (ASIA 2022). Thus, the team began investigating existing clothing designed for the seated body. The team first identified current rain gear sold in stores, such as ponchos and blankets, to better understand how to provide protection for the seated body. One of the designs was a multifunctional garment called the Rayn Jacket, developed in 2015 by Open Style Lab™ and Ryan DeRoche for Betabrand. The jacket was designed to protect bike-riding commuters

Figure 4.49 Quemuel Arroyo wearing qXgo, a bright red rain jacket. Courtesy of Open Style Lab.

and wheelchair users against the rain. The team also researched theoretical methods, such as MOHO—particularly, the OPHI-II (Occupational Performance History Interview-II) Version 2.1 assessment as the underlying foundation toward inclusive design (see chapter 3, MOHO). The MOHO is a client-centered occupational therapy model that examines how volition, habituation (roles and routines), and performance capacity (physical, cognitive, and perceptual abilities) of an individual and their physical/social environment enables occupational participation (Kielhofner 2008). This method helped identify the context of Q's activities and in which situations the garment design would be the most useful.

Observation and Interview

After a field survey of existing products, the team members conducted a semi-structured interview that revealed environmental (physical and social) enablers and barriers to a greater quality of life. Q responded that his current umbrella and GORE-TEX jacket combination fail to protect the gap between his wheelchair back support and wheelchair cushion. This safety risk inspired the team to ideate on a design that has little to no seams to accommodate both Q's work and leisure events.

Figure 4.50 Staci Chan and Kailu Guan discussing with Q his comfort wearing his existing jackets. Photography by Chengcheng Zhao. Courtesy of Open Style Lab.

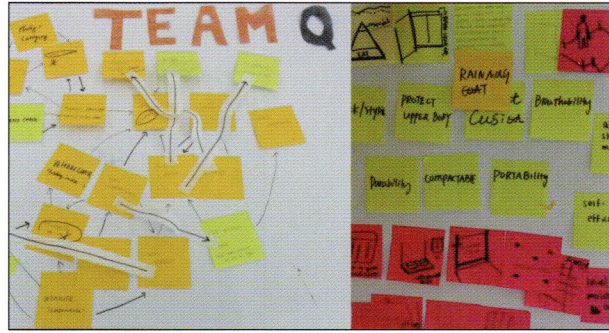

Figure 4.51 Observation notes created by Team qXgo to identify design factors for the jacket. Courtesy of Open Style Lab.

Goal Setting

The team planned to create a garment that (1) protects the wearer's wheelchair cushion from water saturation, (2) protects the seated wearer's skin on their lower extremities that could become infected if wet for too long (The American Occupational Therapy Association 2020), and (3) reduces the need to carry around extra clothing.

Identifying Design Requirements and Challenges

1. Breathable and able to let the wearer independently choose the degree of rain protection, which includes a removable blanket.
2. Portable to let the wearer seamlessly transition from commuting in the rain to work or leisure events.
3. Waterproof throughout the wearer's upper body, trunk, and lap, and protective of their wheelchair cushion.

Prototype Iteration 1

First, the team studied Q's existing outerwear garments to identify better solutions. Q currently sports a Marmot® rain jacket made of GORE-TEX—an ePTFE (expanded polytetrafluoroethylene) membrane designed to be breathable, waterproof, and windproof based on its pore structure (GORE-TEX 2017). Despite

Case Studies, Stories, and Interviews

Figure 4.52a–b Kailu Guan and Chengcheng Zhao testing a muslin prototype design after initial measurements with Q. Courtesy of Open Style Lab.

the jacket's material, Q felt hot when wearing it. Another design issue the team identified with the jacket was the hem—the front was too long, and the back rose up when Q propelled his wheelchair. Q often folded up the bottom half of the front, which made accessing his exterior pockets difficult. Identifying these factors helped the team's first prototype focus on fit and breathability factors for their rain jacket design.

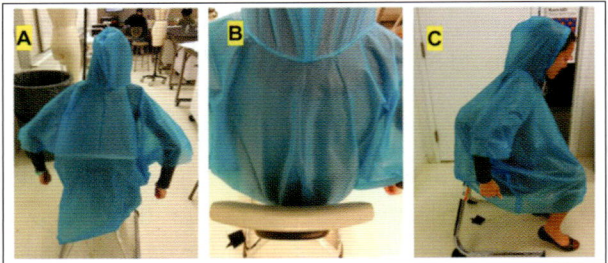

Figure 4.53 Staci Chan wearing blue plastic prototypes of possible jacket silhouettes with varying hem lengths. Courtesy of Open Style Lab.

Fit

The first iteration was not designed as a two-piece. As the team assessed more fittings and iterations, the design became more fitted and eventually became separate pieces. Kailu began taking measurements of Q's body and wheelchair. Because the wheelchair was an extension of Q's environment, the design needed to seamlessly fit his body while not getting caught in the wheels. To address this, qXgo features a shorter front hem that stops at hip level and a longer back hem.

To create an optimal silhouette, the team used existing waterproof materials, such as polyester and plastic, to create shapes for the seated body. Several iterations informed the ideal hem length prior to patternmaking.

To enable maximum mobility of Q's upper body, the team implemented raglan sleeves, a sleeve style that extends to the collar in one piece and

Figure 4.54 Q showing the raglan sleeve design. Photography by Kilian Son for Open Style Lab. Courtesy of Open Style Lab.

features a diagonal seam from the underarm to the neckline. The raincoat also has two layers on the bottom part of the sleeves to prevent abrasion and any rainwater from entering. Because there was a limited amount of room around Q's lower body and the back of his wheelchair, side seams were kept short to prevent the raincoat from gathering at the sides. Extending the back length of the garment also helped further protect Q's wheelchair cushion from the rain, which enhanced his safety secondary to impaired sensation.

Breathability

Because Q often feels hot and sweats despite the breathable elements of his GORE-TEX rain jacket, the team designed a silhouette that could enhance breathability during wheelchair usage. Its features included (1) lining in the sleeves only, (2) a back cut into two different parts—to cover Q's body and the back support of his wheelchair, and (3) double-fold bias tape to cleanly finish all the edges (The American Occupational Therapy Association 2020). The team discovered that a French seam and/or flat-felled seam created fabric bulk that could cause pressure points, which presented a safety risk around the area of Q's injury. Furthermore, the seam and fit of the design eventually informed the team to create a two-piece garment design—a jacket covering the upper body with an apron or blanket-like cover that would protect Q's lap.

Prototype Iteration 2

Portability

Portability is a key factor in the types of garments Q decides to wear. For example, Q does not wear rain pants often because it would require him to carry a second pair to change into. While conducting fittings, the team recognized carrying a separate blanket cover would present another challenge. Should Q desire to take off the blanket, it would need to be stored somewhere on his wheelchair. This prompted the team to focus the next design iterations on exploring a panel cover that could function as both a portable bag for the lap cover and cover the side panels of Q's wheelchair. What resulted were panel covers that also served as a bag to store the blanket when not in use. This would provide a more seamless transition for Q when navigating from outdoor to indoor environments.

Adding a blanket caused design changes to the hem of the jacket. Both pieces had to fit comfortably on Q's waistline. Through trial and error, the team assessed where the top band of the blanket should rest on the body in relation to the front jacket hem. Placing the blanket under the front of the jacket appeared to provide a double layer of rain protection when Q moved. Because

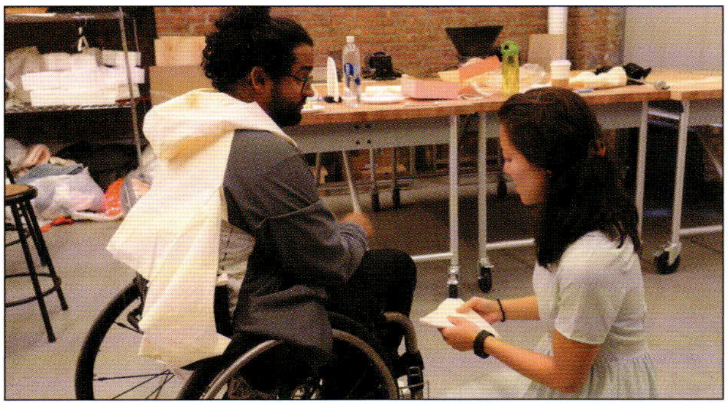

Figure 4.55 Staci Chan assessing the seams and fabric of a muslin prototype with Q. A fitted hood was also constructed in muslin to best support Q's peripheral vision. Courtesy of Open Style Lab.

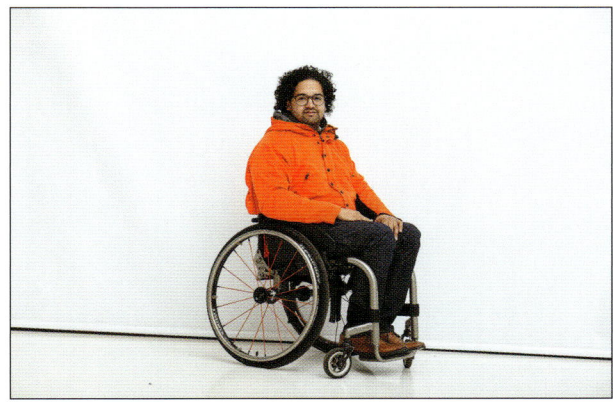

Figure 4.56 Quemuel Arroyo's wheelchair (a custom TiLite ZR frame), which features side panels. Courtesy of Open Style Lab.

Case Studies, Stories, and Interviews   **129**

the front hem was changed, the team also lengthened the back into a curve that met the side seams to cover the space between the wheelchair back support and the wheelchair cushion.

Prototype Iteration 3
Fabrication Tools
The next iterations focused on how the panel covers could attach to the blanket. While designing the lap cover, the team explored ways it could attach to the side panels, such as snaps or clips.

To accomplish this, the team used Rhinocerous® software to create 3D printed clips and wheelchair side panels that were the same shape and size as those on Q's wheelchair (a custom TiLite ZR frame). The first set of clips remained secure when they were pulled in the horizontal axis but were removable from the vertical axis. After several iterations of shape, closure type, and closure placement, the final two side panels were made.

Prototype Iteration 4
To test all the pieces of the design together, Q wore the jacket and lap cover and assembled the panel covers. In doing so, the team learned they needed to refine the shape, closure type, and closure placement for both side panels. The zipper on the panel was repositioned from the bottom into the circumference, creating a more secure fit. This also provided easier access for Q when reaching toward his sides. The team also decided to use magnetic and waterproof FIDLOCK® buckles, as opposed to clip closures. Finally, a brim on the blanket to direct rain flow outwards was also created. The measurements of the brim were determined partially by the type of fabric.

Material Studies
Aesthetic choices were made based on two things: fabric performance and color. A bright orange was picked to provide extra protection for

Figure 4.57 Detachable lap blanket that connects to the side panels. Courtesy of Open Style Lab.

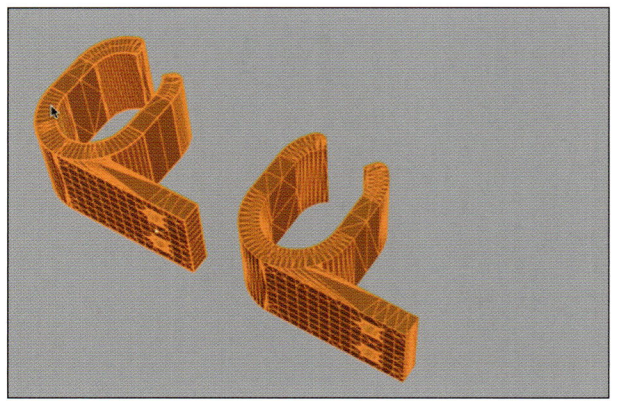

Figure 4.58a–b 3D printed clasps (courtesy of Chengcheng Zhao) designed to attach the blanket to the panel cover. Courtesy of Open Style Lab, Inc.

Figure 4.59 Fabric testing and comparison. Courtesy of Open Style Lab.

Q when he commuted (among cyclists and cars) through the streets of New York. In addition to matching his personal style, the hue also reflected Q's bright, bold, and energetic personality.

The team focused on the textile technology found in Polartec® NeoShell®, Polartec® Power Shield® Pro, and fabrics with cotton and polyester blends in an effort to enhance the garment's rain protection. Using the AATCC Water Repellency Spray Test, results confirmed the waterproof ability of Polartec® NeoShell®. Five meters of the waterproof and breathable fabric were used in the construction of the final product. In addition, the team also utilized high-abrasion fabric to place under Q's elbow.

Project Materials
- Polartec® Power Shield®
- Polartec® NeoShell®
- High-abrasion tent-repair fabric (elbow section)
- 3D printing filament for wheelchair bag closures
- 3D modeling software
- Cotton
- Velcro®
- FIDLOCK® buckles

Final Design Outcome

"When it's raining or snowing, I can feel confident and not worry about not having the right gear. This garment tells the story of who I am: The young, sleek, and sexy guy that I am."

Quemuel 'Q' Arroyo (pers. comm.)

Figure 4.60 Q wearing the final design, including blanket lap cover and top. Photography by Kilian Son for Open Style Lab. Courtesy of Open Style Lab.

Employing participatory design methodology allowed the team to (1) create qXgo with and for Q and (2) positively enrich the design process by combining unique perspectives from design, engineering, and occupational therapy (The American Occupational Therapy Association 2020). The value they placed on open communication, respect, and collective thinking from the first storyboard to the final product contributed to the project's ultimate success—on an individual and systems level.

By applying inclusive design principles to fashion, qXgo holistically considers the wearer's environment in both static and dynamic positions, focusing on breathability, portability, and water repellency while still sporting a sleek look. The raincoat and blanket also invite further explorations into the combination of design, functional textiles, and technology to promote more inclusive experiences.

Case Studies, Stories, and Interviews

How It Works

The stylish raincoat was designed while considering the wheelchair "ecosystem," which includes two side guards. The two-piece product includes the following features:

Top Design
- Back vent for breathability (people with SCI often experience problems with temperature regulation)
- Waterproof fabric (confirmed by the AATCC Water Repellency Spray Test)
- Ultra-bonded seams to prevent water flow
- Fitted hood that allows peripheral vision
- Abrasion-resistant fabric placed under the elbow

Bottom Design
- Detachable lap blanket
- Two wheelchair side guard covers that also serve as storage for the blanket
- 3D printed clips to secure the blanket and side guards

References

ASIA (American Spinal Injury Association). 2022. *Adaptive Clothing for People with Spinal Cord Injury: Where Function Meets Fashion*. Abstract 14. New Orleans, LA.

Bezyak, Jill L., Scott A. Sabella, and Robert H. Gattis. 2017. "Public Transportation: An Investigation of Barriers for People With Disabilities." *Journal of Disability Policy Studies* 28 (1): 52–60. doi:10.1177/1044207317702070 (accessed August 14, 2023).

Christopher & Dana Reeve Foundation. 2023. "What is a Complete vs Incomplete Spinal Cord Injury?". https://www.christopherreeve.org/todays-care/living-with-paralysis/newly-paralyzed/how-is-an-sci-defined-and-what-is-a-complete-vs-incomplete-injury/, February 20 (accessed October 26, 2023).

GORE-TEX. 2017. https://www.gore-tex.com/experience/history (accessed October 26, 2023).

Kielhofner, Gary. 2008. *Model of Human Occupation: Theory and Application*, Fourth Edition. Philadelphia: Lippincott Williams & Wilkins.

The American Occupational Therapy Association. 2020. "Occupational Therapy Practice Framework: Domain and Process Fourth Edition." *American Journal of Occupational Therapy* 74 (2): 1–87. doi:10.5014/ajot.2020.74S2001 (accessed August 14, 2023).

Wong, Sandy. 2018. "Traveling with Blindness: A Qualitative Space-Time Approach to Understanding Visual Impairment and Urban Mobility." *Health & Place* 49: 85–92. doi:10.1016/j.healthplace.2017.11.009 (accessed August 14, 2023).

Wong, Sandy, Sara L. McLafferty, Arrianna M. Planey, and Valerie A. Preston. 2020. "Disability, Wages, and Commuting in New York." *Journal of Transport Geography* 87. doi:10.1016/j.jtrangeo.2020.102818 (accessed August 14, 2023).

Swipe
Christina Mallon, Julia Liao

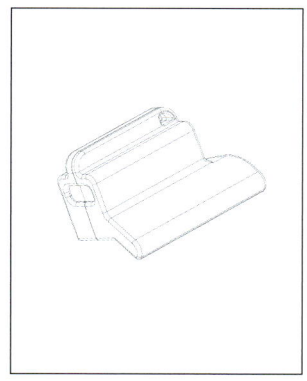 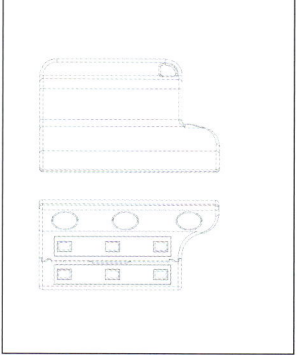

Figure 4.61a–b Vector illustrations of Swipe created by Julia Liao. Courtesy of Open Style Lab.

> **Key Functional Features**
> Portability
> Easier reach
> Weight

Summary
Swipe is a 3D printed assistive device worn around the neck that helps any wearer swipe a MetroCard in the New York City Subway entrance.

Design Goal
To create a wearable device that helps its user swipe a MetroCard independently and hands-free.

Team Swipe
Christina Mallon: Client, subject-matter expert, and inclusive designer
Julia Liao: Product designer

Collaborator Acknowledgments
Parsons School of Design, Open Style Lab™, Ultimaker, Estee Bruno, Claudia Poh

Background
Products that aid the population in maintaining their independence can often carry the assumed stigmas of medical devices and assistive equipment (Resnik 2009). Subways are one of the most essential public transportation methods people use to navigate New York City. In 2017, the MetroCard system still used a plastic card and metal reader, presenting a challenge for many people who must swipe a plastic card at a particular height, speed, and grip.

The following design process investigates a possible wearable solution that addresses subway transportation experiences that are not accessible to people living with disabilities. The result was Swipe—a combination of AT and product design produced as a wearable device. Without the use of technology, everyone getting older or facing a disability may choose to employ

Figure 4.62 Christina Mallon wearing Swipe to access a New York subway entrance. Courtesy of Julia Liao.

a personal caregiver, move to an assisted living facility, or seek ways to hack inaccessible designs. "We (PWD) have to figure out new and innovative ways to do things other people take for granted: eat, get dressed, and brush our teeth. We're the original hackers," writes Christina Mallon for *Fast Company* (2018).

Introduction

As mentioned in the Unparalleled case study, Christina has a rare form of ALS that has left her arms and hands completely paralyzed, making dressing and traveling difficult. One of her biggest challenges is using the New York City Subway system to get to and from work. Its heavy steel doors, turnstiles with sliding card readers, and bustling rush hour foot traffic often force Christina to ask someone for help. So, she and Julia set out to create something functional and stylish that could ease her daily commute.

Design Process

1. Observation
2. Goal Setting
3. Prototype Iterations

Observation

"I'm disabled by my environment, so I find ways to hack it."

    Christina Mallon (pers. comm.)

Julia visited Christina at her home on several occasions to see the different ways she made her environment accessible. They included installing a detachable handle on the refrigerator door, mounting hooks on the wall in her living room, placing objects and appliances lower to the ground, and using speech-controlled devices, like Amazon Alexa and the speech-recognition program Dragon. Julia also observed how Christina uses her feet to maneuver her phone—but always with her own unique hack: attaching a keyring and flexible cord to lower it to the floor. As mentioned in Unparalleled, Julia noticed that Christina independently dresses by laying out a garment on her furniture before swooping her head through the neck hole to put it on. The way Christina used her neck was a key observation that later informed the final design.

Goal Setting

It was apparent to Julia that creating something with Christina would benefit others, too. "A person with a disability, injury, or symptoms of aging might need better access and more frequent donning and doffing due to having urinary catheters, prosthetics, diapers, or orthopedic casts attached to their bodies. Being able to dress yourself is not only about physical ability but also having control over your own privacy and independence," says Julia (Julia Liao, pers. comm. 2018). Together, they identified three necessary design requirements:

1. Able to be worn independently with little motion.
2. Provides an easier way to enter subway stations.
3. Comfortable and lightweight enough to be worn around the neck for long periods of time.

Prototype Iteration 1

On one occasion, Julia observed how Christina commutes to work: Every morning, she's visited by a nurse who helps her get ready for the day before walking her to the subway and swiping her MetroCard. While this arranged system has helped her tremendously, it's limiting when she wants to meet up with friends or simply spend a day enjoying the city. To increase Christina's independent participation in social activities such as these, she and Julia began exploring the use of wearable products—specifically, something that could hold a MetroCard.

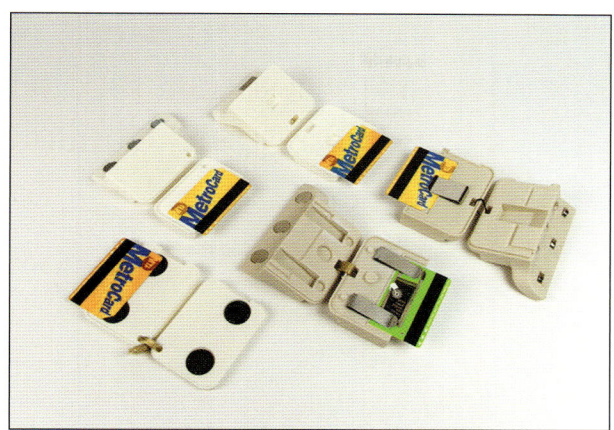

Figure 4.63 Design iterations of Swipe using 3D printing. Courtesy of Open Style Lab.

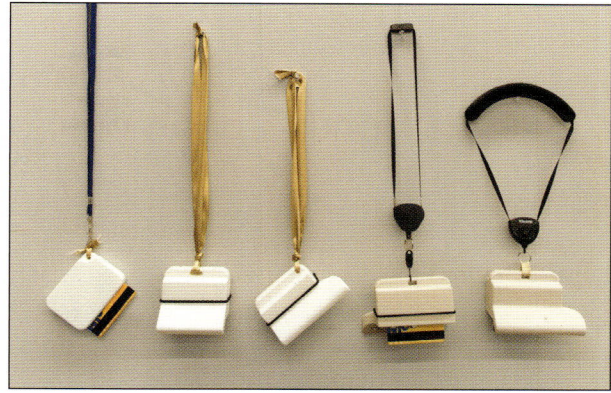

Figure 4.64 Neck straps of different materials and lengths attached to Swipe design iterations. Courtesy of Open Style Lab.

With the help of a desktop 3D printer, Julia made several iterations of a MetroCard holder that she then assembled by hand. The many benefits of using a 3D printer include its accuracy, price point, and ability to easily produce multiple quantities on a much smaller scale than injection molding in traditional manufacturing. They can also accommodate various plastic filaments, which was essential for creating Christina's assistive device. Swipe was printed using PLA (polylactic acid) and PE (polyester) plastic filaments with PVA (polyvinyl alcohol) dissolvable support material. This method allows certain cavities in the model to be smoother and more accurate to ease the process of assembling pre-made parts, such as magnets, ball bearings, and clips.

Prototype Iteration 2
Julia's second iteration featured a rim for Christina to bite onto before using the prototype; however, they both quickly realized it was too difficult to maneuver with one's mouth. So, they decided to leverage her body motion instead, which would ultimately be more comfortable and less noticeable to passersby. An initial challenge was creating a design that provided enough structural support for Christina to swing the device onto the card reader. Because the MetroCard holder wouldn't latch, Julia added magnets to the bottom, as well as a fabric necklace that it would hang from.

Prototype Iteration 3
While the magnets did help guide the MetroCard holder into the subway turnstile, their strength turned out to be a problem—either causing the prototype to swing too fast or too slow. Julia identified the best magnetic fit, strength, and speed that would suit Christina.

Prototype Iteration 4
The last design iteration focused on Christina's ability to don and doff the MetroCard holder. Because its initial fabric necklace attachment wasn't adjustable, Julia added a retractable lanyard with a button that Christina could easily press onto with her body. She could also stretch it to take it off faster, causing it to retract and bring the device upwards. This helps widen the entry point around her neck and adjust the length accordingly. Finally, a soft, bendable NinjaFlex® TPU 3D printing material was added to the back of the neckline to help Christina put on her device.

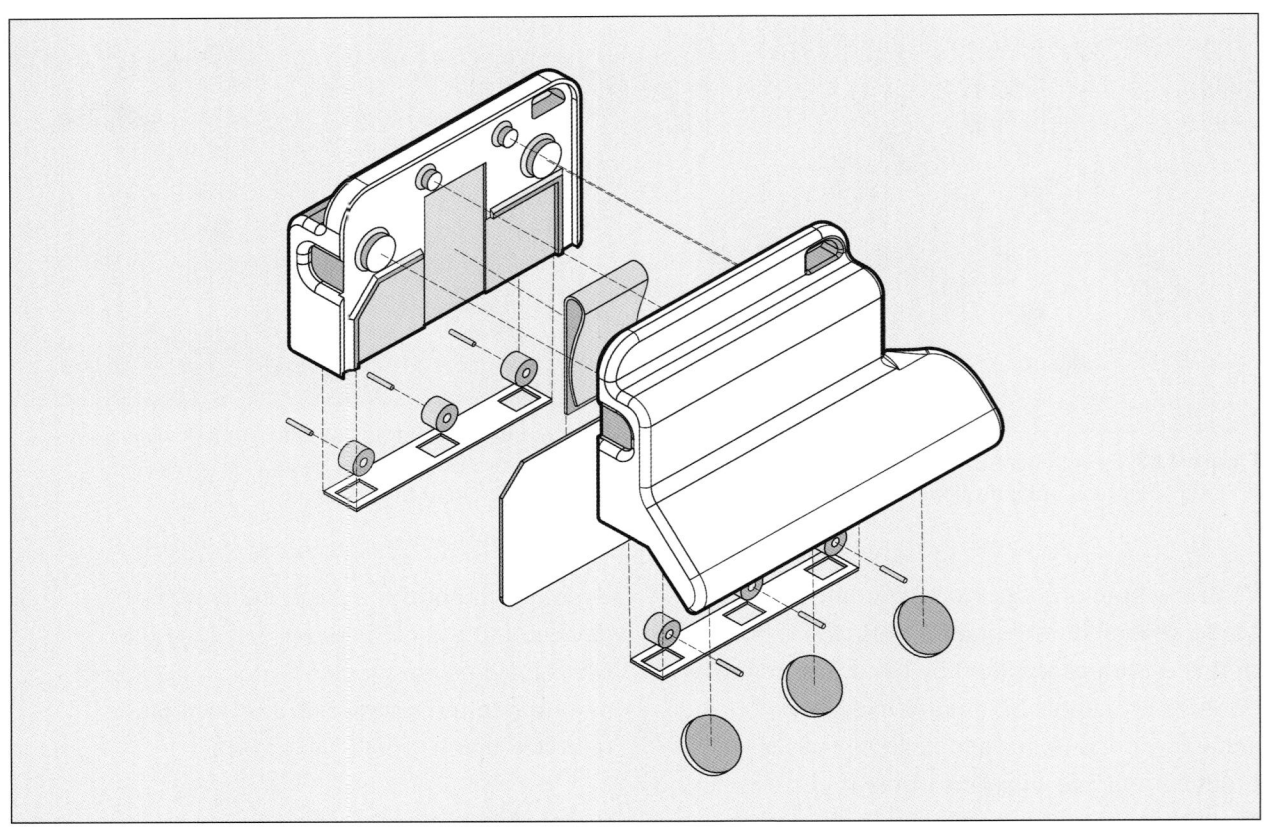

Figure 4.65 Final CAD rendering of Swipe design featuring the magnets and ball bearings. Courtesy of Open Style Lab.

Materials
- PLA and PE plastic filaments
- Desktop 3D printer
- Retractable lanyard

Final Design Outcome

Swipe, the wearable product, has stationary tension fit clips in the body that hold a MetroCard flat and in place. It also features magnets and ball bearings that help navigate it onto a New York City MetroCard reader. The wearer then needs to use their range of motion to align the MetroCard holder's outer ledge before nudging, pushing, or pulling it through in a linear swiping motion. In addition to helping Christina live a more independent life, Swipe has the potential to assist anyone living with ALS, MS, and arthritis.

References

Core Jr. 2018. "Meet Julia Liao, the Winner of This Year's Core77 x A/D/O Residency." *Core77*, March 19. https://www.core77.com/posts/74996/Meet-Julia-Liao-the-Winner-of-This-Years-Core77-x-ADO-Residency (accessed August 14, 2023).

Resnik, Linda, Susan Allen, Deborah Isenstadt, Melanie Wasserman, and Lisa Iezzoni. 2009. "Perspectives on Use of Mobility Aids in a Diverse Population of Seniors: Implications for Intervention." *Disability and Health Journal* 2 (2): 77–85. doi:10.1016/j.dhjo.2008.12.002 (accessed August 14, 2023).

Mallon, Christina. 2018. "The Most Valuable Person You Haven't Hired Yet." *Fast Company*, July 9. https://www.fastcompany.com/90180550/the-most-valuable-person-you-havent-hired-yet (accessed August 14, 2023).

Interview
Human Factors and Occupational Therapy— Michael Tranquilli

Figure 4.66 Photograph of Michael Tranquilli taken by Kilian Son. Courtesy of Kilian Son.

Michael Tranquilli is an occupational therapist and adjunct faculty member in the Department of Occupational Therapy at New York University.

Grace
What's your background, role, and field of practice?

Michael
I'm a private practice occupational therapist providing rehabilitation services to individuals with neurologic, orthopedic, and musculoskeletal impairments and disorders. I provide consulting services in the fields of workplace ergonomics, universal design, aging in place, and ADA compliance. I'm also an adjunct faculty member in the Department of Occupational Therapy at New York University, where I educate students in adaptive, compensatory, and rehabilitative strategies for physical disabilities.

Grace
Could you expand on workplace ergonomics, universal design, and how we can think about the body in different situations?

Michael
Workplace ergonomics is a discipline focused on understanding and improving the interaction between workers, their tools or tasks, and the work environment. Universal design is an interdisciplinary approach focused on creating environments, products, and services that accommodate the widest range of human abilities. Upon comparison, universal design is more broadly concerned with accessibility, whereas ergonomics is more specifically concerned with human performance. Together, these two fields of study measure the complex variations in human factors and contextual demands to enhance the body's performance in different situations.

Grace
Could you share any experiences about what dressing at home might mean? From an occupational therapy lens, we've both seen people sometimes get dressed lying down, dressed standing up, or have assistance in dressing.

Michael
Dressing is a complex activity that can be physically and cognitively demanding. Since dressing can occur on different surfaces and in multiple environments and positions, careful consideration is placed on ensuring the safest and most functional performance with the least amount of energy consumption. Successful or independent dressing in the home environment encourages meaningful participation in social and professional activities.

Grace
Many people are seated in the workplace for long hours, like some wheelchair users. Could you expand more on people's dress situations or body posture?

Michael
Yes. Body posture and dress can have a negative effect on health. For example, increased rates of low back pain and injury occur when people are seated for prolonged periods. Since clothing is typically measured with the standing body in mind, clothing has a propensity to bunch up at the hips and behind the knees when seated, trapping heat and moisture and exposing the skin to abrasion and pressure sores. Additionally, the buttocks, hips, and thighs are wider when seated compared to a standing position. Clothing measurements need to reflect these differences in body positioning to improve comfort and functionality.

Grace
Occupational therapy offers a broader way to think about clothes and getting dressed. Are there key learning moments for designers during the making process on how their decisions might affect the body?

Michael
Yes. Key learning moments can occur through success and failure. Along with the individual's abilities and limitations, successful designers consider the surfaces, settings, materials, and closures used during the making process. Failure to assimilate these factors during clothing design will restrict motivation and performance in dressing. Ultimately, the learning process must never supersede the safety and well-being of the intended user.

Grace
What preventions in design are overlooked by designers? How can designers better learn how their choices could impact someone's health?

Michael
As occupational therapists may not understand all the complexities of fabrics, patterns, and stitching, fashion designers may not understand all the complexities of injuries, impairments, or disability. When designing for the aging or disabled populations, occupational therapists and designers must not overlook the time, energy, or assistance required to complete the dressing process. Any clothing design or process that increases the time, energy, or assistance required will most likely have a negative impact on performance and health. Most importantly, patience and communication with the user are required to fully comprehend how their personal experiences can bring meaningful impact to the design process.

Grace
Are there any techniques or thought processes that you'd like to share with students?

Michael
I think it's important for students to know that occupational therapists are the original healthcare hackers. We organize and design processes and systems that facilitate the successful completion of tasks based on a user's strengths and limitations. Students

should always embrace a collaborative approach to learning when attempting to meet the diverse needs of persons with disabilities. Occupational therapists, for example, are very proficient at taking body measurements of persons in a seated position or with asymmetrical postures. Designers would greatly benefit from learning these techniques to ensure proper fit, comfort, and functionality when creating products for atypical body types.

Grace
I've seen so many examples around grip and dexterity. Are there certain concepts that would be helpful for designers to keep in mind when the body comes in contact with clothing closures?

Michael
Understanding both the cognitive limitations and physical abilities of the user's interaction is a critical aspect of determining the most efficient approach to clothing closures. For example, a person may comprehend the process of buttoning but may not possess the grasp to hold or dexterity to manipulate the button through the hole. Conversely, a person may possess the dexterity to tie a knot but is incapable of appropriately sequencing the steps. Both scenarios represent a failure in design and/or process. Clothing designers must keep in mind that the location of closures should be positioned for accessibility and that the physical characteristics of closures should be compatible with the user's abilities. Matching the physical characteristics and process of closure with the users' abilities is the best design approach.

Grace
Everyone has a different approach to dressing, like the way we dress or the sequence of dressing. Could you expand upon people's habits or memories of getting dressed in a specific way, perhaps that is more comfortable for them?

Michael
Dressing is a very personal experience. The process and sequence of dressing can be extremely individualistic or stereotypical. Although most behaviors or habits arise from personal strengths or weaknesses, people tend to prioritize their process according to factors of simplicity, efficiency, intuition, and safety, which can vary depending on the type and style of clothing, the environment, or the time of day in which dressing occurs.

Grace
Right, it can be for all people. Especially in the aging population?

Michael
Yes. The world's population is aging, and aging exposes people to a greater risk of impairment and disability. We should all aspire to solve the problems that diminish our quality of life throughout our entire lifespan.

Grace
We discussed a lot about performance and functionality, but there's that aspect of dignity or emotional aspect in design too. The sense of accomplishment that one feels when a client can either do things independently or with less assistance. How have you seen dressing as an activity, within the lens of occupational therapy, as a fulfilling accomplishment for people?

Michael
Yes, that's a great question. Occupational therapists have a diverse knowledge base, including backgrounds in psychology and sociology. Specifically with dressing, our goal is to enhance a person's self-efficacy by minimizing or overcoming barriers to successful completion. Self-efficacy refers to a

person's belief in their ability to execute behaviors that produce specific performance outcomes. A person with high self-efficacy strongly believes they can accomplish the steps required to complete the activity and is, therefore, more likely to engage in an activity. Frequent engagement in an activity typically results in performance improvement over time and increases confidence in one's own ability or self-esteem. Through the lens of an occupational therapist, there is nothing more rewarding than witnessing the pride achieved when people become proficient or independent in activities of daily living.

Conversation Points
- What kind of collaborators in the health and therapy industry would you need for your projects?
- What are the complexities of the environment that enhance or inhibit a person's performance, such as getting dressed?
- How can you apply ADLs to projects that are designing with and for disability?
- What is a person's body posture over the course of a day?
- Have you encountered dressing challenges yourself?

Zipback Jacket
Justin Moy, Prow Samsethsiri, Yuchen Zhang, Priyal Parikh, Ruthie Merrell

Summary
The Zipback Jacket is a multi-season garment with a unique structure and design that maximizes comfort for an active high schooler with muscular dystrophy and eases the process of donning and doffing for caretakers and dressing assistants.

Figure 4.67 Team Zipback Jacket from left to right: Ruthie Merrell, Prow Samsethsiri, Justin's sister, Justin Moy, Priyal Parikh, and Yuchen Zhang. Courtesy of Open Style Lab.

Key Functional Features
Protection
Thermoregulation
Comfort
Attention signaling

Design Goal
To create wind and waterproof outerwear that's breathable, temperature controlled, and suitable for a wheelchair user.

Team Zipback Jacket
Justin Moy: Client, subject-matter expert, and disability rights activist
Prow Samsethsiri: Client's mother and caretaker
Yuchen Zhang: Designer
Priyal Parikh: Designer and Engineer
Ruthie Merrell: Occupational therapist

Background
According to the MDA (Muscular Dystrophy Association), muscular dystrophy causes progressive weakness and loss of muscle mass, which affects 2–3 percent of the US population in varying degrees. Over time, it makes standing up straight difficult and often causes a secondary condition called scoliosis, a sideways curvature of the spine. Despite such physical changes, clothing has not yet provided options that are aesthetically pleasing and functional for a range of body postures. This project explores how to create an outdoor jacket for the posture of a seated person and the curvature of the spine.

Introduction
Justin is a bright, creative high school student with muscular dystrophy who's actively involved in musicals, singing groups, and the Model UN. He aspires to be a biochemist and spends much of his free time participating in the MDA's camps and telethons. Justin's home had been renovated for accessible living to better accommodate the use of a wheelchair. For example, elevators were installed so that Justin could access his room on the second floor, and lift tracks were added to the ceiling to move him between his bed and toilet. While his home is becoming more accessible, his wardrobe options are not.

Case Studies, Stories, and Interviews

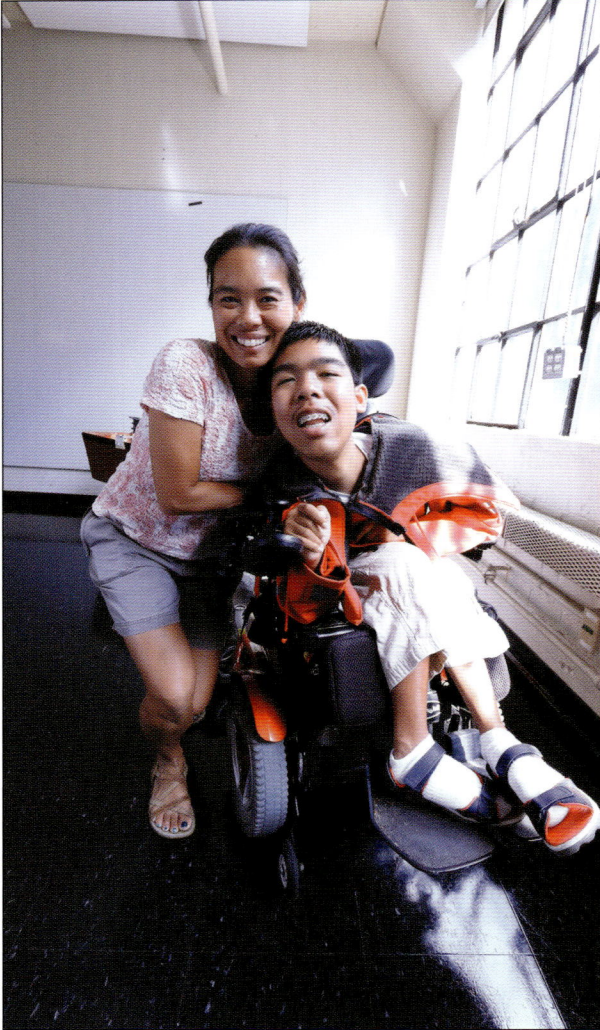

Figure 4.68 Prow Samsethsiri and Justin Moy wearing Zipback Jacket. Courtesy of Open Style Lab.

The Project
There are three main components of this design process:
1. **Patterning:** Comprehensive measurements were taken of Justin's entire body to create unique patterns that fit him in ways off-the-rack clothing couldn't—adjusted for his posture and the curvature of his spine.
2. **Construction:** A zipper was placed in the center back of the garment to help Justin's mom with supportive dressing. Additionally, the jacket was made shorter in the front to prevent the fabric from bunching in his stomach area and trapping heat.
3. **Materials:** Performance fabric was used for protection against cold-weather elements; silicone was screen-printed onto the back of the jacket to prevent the hem from riding up while being worn in a wheelchair; and bonded text was added for personalization and increased visibility on the sleeve of the jacket.

Design Process
1. Research
2. Observation and Interview
3. Goal Setting
4. Prototype Planning and Development

Research
As his primary caretaker, Justin's mother, Prow, helps him dress every day. The team, therefore, focused on creating a jacket that Justin and Prow could quickly don and doff with ease. They brainstormed various scenarios (going to school, skiing, and camping in weather ranging from 45°F to 70°F) and researched clothing patterns for different body stances. That led them to notice that Justin's posture was similar to a cyclist's body position, prompting them to research further into performance wear for athletes in similar body positions and contours to Justin's.

Observation and Interview
Other than desk research, site visits were an invaluable way for other team members to better understand Justin's needs and wants. Members of the team were able to see the accessible additions to his home through a site visit and learn more about his fashion style. Justin was very clear about his preferences. For example, he

favored khaki or flannel designs and garments with fleece. This informed the team to search for materials and colors that best complemented Justin's existing style preferences.

Goal Setting

After the site visits, the team identified four very specific goals. First, the garment's design had to avoid fabric bunching in the stomach area. Second, the design needed to regulate body temperature. Contractions in Justin's neck, elbows, and hip caused his torso to bend forward, his forearms to fold, and his head to lean backward—this meant excess fabric would gather, trap heat, and make him feel uncomfortable. Design factors that provided Justin more independent control over parts of the garment were important, so the design needed to fit his range of motion and body contour. Third, the jacket had to be long enough to cover his entire back. Lastly, Justin also needed a hood to protect his face, ears, and neck from the cold.

Prototype Planning and Development

Prototype Iteration 1

Before creating the actual jacket, team members wanted to better understand Justin and Prow's comfort levels with a zipper, asymmetrical lengths for the front and back of a garment, and magnetic seams. To accomplish this, they first used a white T-shirt (Prototype 1A) and a white button-down shirt (Prototype 1B) to test out various closures and hem lengths.

Prototype 1A was cut to be shorter in the front and longer in the back with a zipper. While the front length significantly reduced fabric from bunching around Justin's stomach area, the zipper was difficult to manipulate independently. The length helped Prow and Justin to don and doff easier, but Justin thought it made the shirt also look like a hospital gown. This insight was important for the team to strongly consider when working on the silhouette of their designs, wanting to avoid any resemblance to a medical garment.

Building off the 1A, Prototype 1B featured magnets sewn into both sleeves: in the inseam under the right arm and on the outside between the forearm and bicep on the left arm. While the magnets on the left arm made it difficult for Prow to guide Justin's hand through the sleeve, they both decided the magnets on the right arm were a good method for temperature control. This iteration provided insight into the ideal sleeve silhouette that would best fit Justin's needs and style.

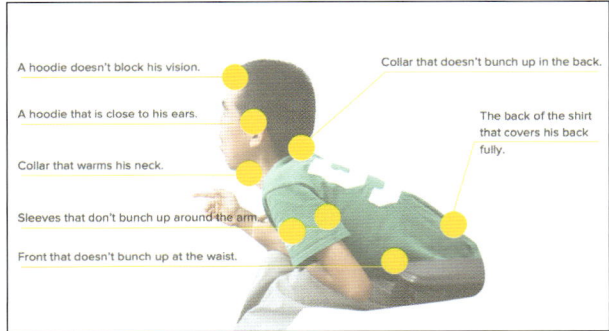

Figure 4.69 Diagram of identifying design goals. Courtesy of Open Style Lab.

Figure 4.70 Two sleeve designs in muslin with black spandex designed by Yuchen Zhang. The top sleeve design has a spandex fabric that wraps around the entire arm, while the bottom sleeve has spandex only on the elbow area. Courtesy of Open Style Lab.

The team heavily relied on Justin's exact measurements. This step was extremely challenging since there were no body forms or reference models the team could use. Members traced Justin's body onto a large piece of cardboard and marked areas that had little to no range of motion, such as his back. During the initial stages of prototyping and iterative design, the team members wanted to explore 3D modeling tools that could visualize Justin's body and inform construction and fit.

The goal was to create a new mannequin style to use for draping. However, isolating Justin from his chair proved challenging for the body scanner, which couldn't subtract the assistive device from his body. Considering the technical challenges faced, the team discarded the idea of using 3D modeling—highlighting the lack of accessible body forms for designers to reference and use. However, Justin's availability to test prototypes eliminated the need for an accurate mannequin. Ultimately, the prototypes were created using a regular mannequin and tested weekly by Justin. While struggling to create an adequate mannequin, more examples of Justin's body structure were researched through discussions with Prow.

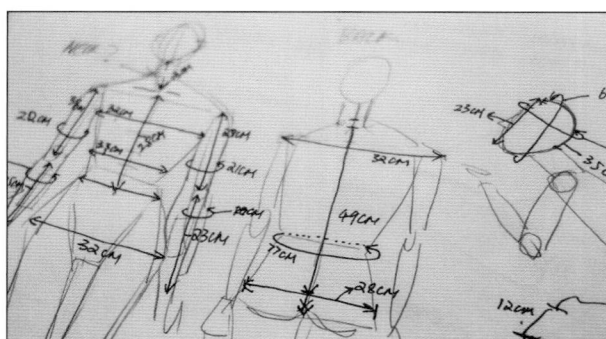

Figure 4.71 A sketch of Justin's measurements in comparison to standing upright body forms in similar clothing sizes, drawn by Yuchen Zhang. Courtesy of Open Style Lab.

Prototype Iteration 2

Prototype 2 was a garment created from muslin, a lightweight plain-weave cotton, that was donned from the front and closed in the back. The inner (left) flap featured a Velcro® band that closed at Justin's waist, and the outer (right) flap had button clasps that connected to the left shoulder. Both sleeves also had different styles—one had stretchy fabric on the back from the mid-forearm to the mid-upper arm, and the other had stretchy fabric around the elbow. The overall design was intended to test the act of donning and doffing the item, as well as how much it bunched around the arms.

Prow found the design easy when helping to get Justin dressed. Prow and Justin agreed that the sleeve with the stretchy fabric between the mid-forearm and the mid-upper arm was easier to pull up; she had more material to grab onto, and he felt it was less constricting (see fig. 4.72). However, the opening of the sleeve prototype was too small for Justin's hands, which formed a

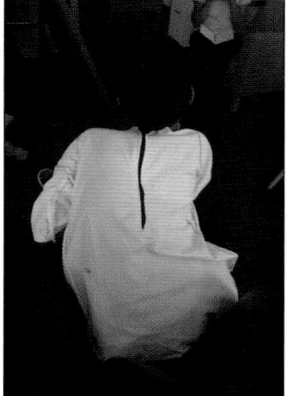

Figure 4.72a–b A photo of Justin wearing a shirt prototype with a zipper in the back to check for neck fit. Two shirt design prototypes for front design and back design with zipper placement. Courtesy of Open Style Lab.

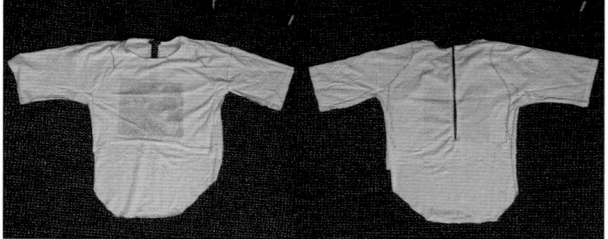

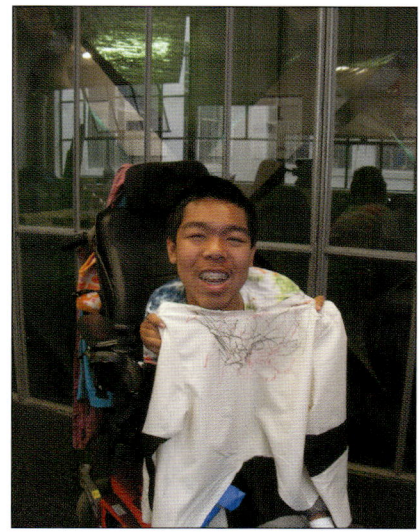

Figure 4.73 Photo of Justin holding a shirt prototype design with pen marks that indicated easy-to-reach spots for a zipper. Courtesy of Open Style Lab.

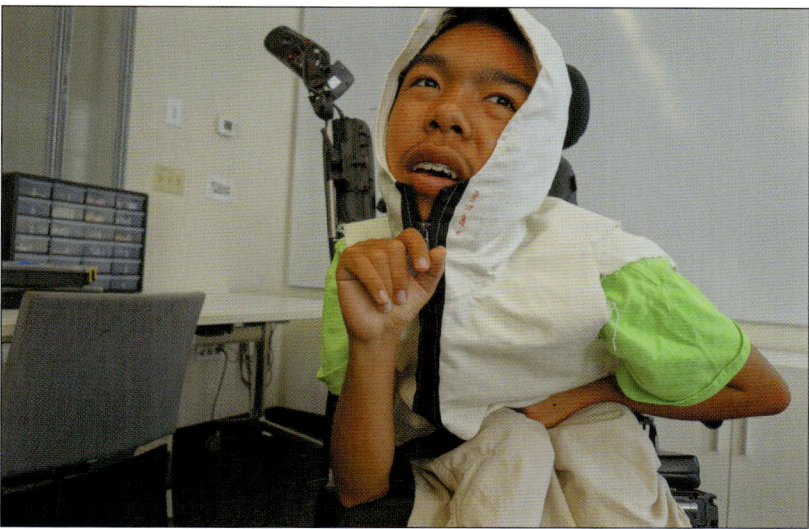

Figure 4.74 Justin testing out a zipper hoodie prototype design in muslin fabric. Courtesy of Open Style Lab.

tight fist. As a result, the team added expandable cuffs with zippers and Velcro® belts, which also helped regulate temperature and provided Justin the option to adjust the sleeve opening.

After the jacket pattern was decided upon, Yuchen, Priyal, and Ruthie focused on pinpointing Justin's precise range of motion. They were aiming to create something Justin could independently manipulate himself. To accomplish this, Justin used a red pen with his left hand and a black pen with his right to draw on the muslin wherever he could reach. The pen marks helped visualize where a closure like a zipper should be best placed.

Prototype Iteration 3
As the design process continued, Prototype 3A was created and modified to reflect the team's discoveries and Justin's preferences. For example, Justin preferred a jacket to function as a wind and waterproof garment and have a thin lining for warmth. Prototype 3A also experimented with a drawstring loop near Justin's neck, but the placement of the drawstring required Justin to spend much more energy. Based on this iteration, Prototype 3B incorporated a different drawstring placement, and the team researched various materials. In addition, Prototype 3B was designed to test for a zipper placement on the upper right side of Justin's chest. The team identified an ideal placement for the zipper design that would be within Justin's reach. After assessing both designs, Yuchen, Priyal, and Ruthie worked with

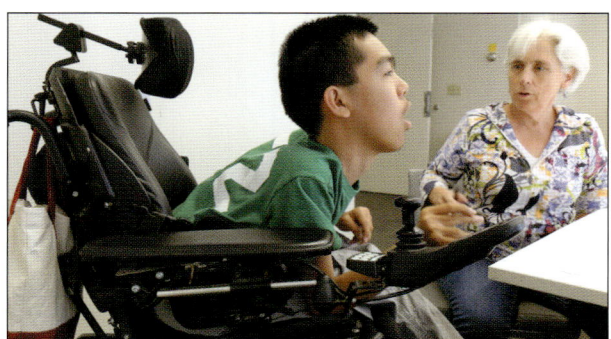

Figure 4.75 Ruthie Merrell discussing with Justin Moy priorities of wants and needs when considering fabric choices. Courtesy of Open Style Lab.

Case Studies, Stories, and Interviews

Justin and Prow to prioritize what details were most important to inform material research. They chose performance fabrics that were wind-resistive yet light in weight created by Polartec®.

Prototype Iteration 4
Next, the team combined Prototype 3A and 3B to create Prototype 4: the neck and shoulder were modified based on the measurements taken. In doing so, the team discovered a manual zipper was difficult for Justin to use. They also continued material research and found prototyping in canvas instead of muslin better depicted a well-fitted hood with a magnetic zipper in the front. The sturdiness of the canvas allowed for a more accurate representation of the type of fabric weight the team would need.

Justin preferred a hood design that would fit around his ears and chin without a zipper being too close to his mouth (see fig. 4.76a). A cloth tag on the zipper helped Justin find it more easily and pull it quickly. The team decided to add the cloth tag and discovered it was more accessible to move the drawstring from the side of the jacket to the front. Other changes included shifting the inner back flap fastener closer to the shoulder than the waist, adding sleeves, and increasing the overall size to accommodate fleece underneath.

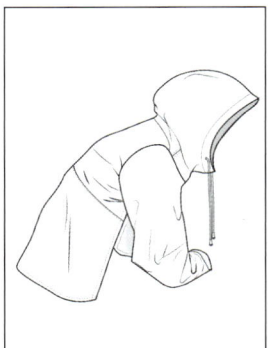

Figure 4.76a–b Illustrations of drawstrings and final silhouette of jacket design created by Yuchen Zhang. Courtesy of Open Style Lab.

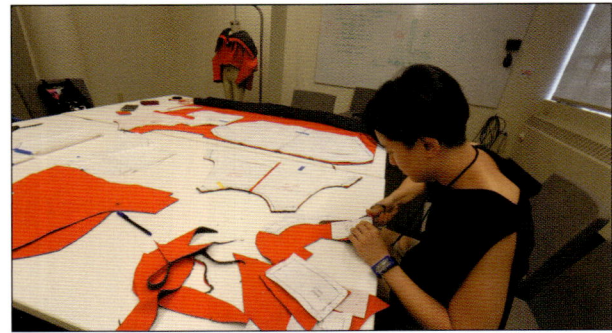

Figure 4.77 Designer Yuchen Zhang cutting final fabrics from pattern designs. Courtesy of Open Style Lab.

Prototype Iteration 5
Prototype 5 proved to be confusing for people to use, particularly for Prow. The double fastening was uncomfortable and restricting for Justin. Major changes were made for Prototype 6: the back flap was discarded; the original idea of using a zipper in the back was implemented (in addition to the front zipper) from the top of the hood to his waist; the sleeves were modified to be wider at the elbow and shaped like a slightly folded arm; and Velcro® was attached to the cuffs to control temperature regulation and the process of donning and doffing.

Final Design Outcome
The Zipback Jacket features a zipper in the center back from the edge of the hood, which makes it easier for Prow when helping Justin with his sleeves. The garment is shorter in the front, long enough to cover his back, and adjusted for Justin's posture and spine curvature. It is made with Polartec® High Loft™ and Polartec® Power Shield® Pro for weather protection, has a silicone screen printed onto the back to prevent it from riding up, and includes bonded text for personalization and increased visibility.

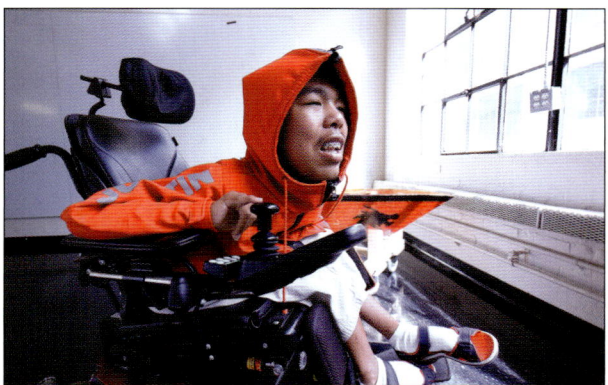

Figure 4.78 Justin wearing a red jacket design, Zipback Jacket. Photography taken by Grace Jun.
Courtesy of Open Style Lab.

Figure 4.79 Design details and closure up of the back of the jacket. Photography taken by Alex Tosti.
Courtesy of Open Style Lab.

Project Outcome

Justin once said in an interview with Region 8 News, "No matter your problem, there is always a creative way to solve that problem." (2016). This positivity guided the design process and was reflected in the final product: a bright red jacket with "Justin" in silver on the sleeve. The Zipback Jacket supports Justin to continue engaging in various activities year-round, express his unique style, exercise his independence, and easily get dressed with assistance from his mother.

References

Region 8 News. 2016. "Fashion w/ Special Needs." September 19, 2016. Video, 0:30. https://es-la.facebook.com/Region8News/videos/10153877266253148/

Ease
Eliza Mury, Aimee Mury, Uma Desai, Chrissy Glover, Elizabeth Riley

Figure 4.80 Members of Team Ease from left to right: Chrissy Glover, Uma Desai, Elizabeth Riley, and Eliza Mury. Courtesy of Open Style Lab.

Key Functional Features
Durability
Comfort
Compression

Summary
Ease is a line of durable, seamless, and stylish T-shirts created with and for people experiencing skin sensitivity.

Design Goal
To create a long-lasting, comfortable top for Eliza that enhances her activity level and daily routine.

Team Ease
Eliza Mury: Client and subject-matter expert
Aimee Mury: Eliza's mother and advisor
Uma Desai: Engineer
Chrissy Glover: Fashion designer
Elizabeth Riley: Occupational therapist

Introduction
Eliza is an active young girl who enjoys gymnastics, swimming, arts and crafts, and difficult 300-piece puzzles. Aimee describes Eliza as having ASD (autism spectrum disorder) and OCD (obsessive-compulsive disorder). Eliza has sensory preferences that favor proprioceptive input, necessitating that clothing fit tightly around her body. Eliza often picks at the seams and hems of her clothes until they are torn completely apart. As a result, Eliza's new clothes quickly become unwearable, with a lifespan of just one minute to one month. The goal of this collaboration was to create a clothing solution that would be durable enough to stand up to Eliza's picking habit while meeting her sensory needs.

Design Process
1. Research
2. Observation and Interview
3. Goal Setting
4. Prototype Planning and Development

Research
Tactile sensory preferences were critical to the process and parameters for the shirt design. Initial research was referenced from a 2007 study by Dunn and Tomchek that helped inform team members about the autism spectrum and Winnie Dunn's *Sensory Profile* (1999) evaluations for children. The Sensory Profile is an evaluation tool used in occupational therapy to establish a child's sensory processing, which is the way one's brain interprets input from the six senses: touch, taste, smell, hearing, vision, and proprioception (Dunn

1999). Proprioception is the ability to recognize where one's body and joints are in space. This evaluation tool is based on the interconnected neuroscience and sensory integration approaches. Because the Sensory Profile (Dunn 1999) identifies a child's perception and preferences, it was a useful tool for establishing Eliza's sensory preferences, specifically her persona, preferred activities, and possible pain points in clothing. This tool was administered qualitatively to Eliza and her mom, and the results of using such a tool were shared with other members of the team. In the end, Eliza's specific sensory input needs were carefully considered, and the team decided to focus on shirt designs. Due to Eliza's tendency to rip and tear fabric and seams, the design of the shirt needed to be extremely durable.

Observation and Interview

The team observed Eliza's pain points were primarily her sensory preferences: She prefers a tight, compressive fit that fulfills her desire for proprioceptive feedback to interact successfully with her environment. Eliza's tactile preferences are selective and clear, preferring soft fabrics without seams and tags.

Goal Setting

After discussion and initial observations, the team's design requirements for the fit of the shirt design were identified. The shirt, Ease, needed to have limited and flat seams without threads or bulky hems to eliminate irritating tactile input. In addition, it needed to be durable and provide a tight, moderately compressive fit to provide proprioceptive feedback. The design also needed to be personalized and made to look and feel special.

Figure 4.81 Shirt design sketch drawn by Chrissy Glover. The shirt, Ease, features a raglan sleeve design and no side seams. Courtesy of Open Style Lab.

Prototype Planning and Development

Initially, adding interactive components to the design was considered. This idea was an attempt to provide Eliza with an alternative stimulus to capture her attention, thus preventing and redirecting her ripping behavior. So, the team tested various materials with Eliza—including Velcro®, elastic, vinyl, leather, 3D printed plastic, and fabric printed with thermochromic dye. The team experimented with a sewn-on Velcro® strip and flap that could be pulled on and off to gauge Eliza's interest. Eliza didn't show interest in playing with the Velcro® the way it was intended, instead ripping it completely off the shirt. With this result, interactivity proved a nuisance to Eliza and her caretaker, so it was dismissed as an option when prototyping a shirt. Members of the team also learned about Eliza's preference for fitted clothing, which inspired the group to research more into smooth synthetic stretch fabrics.

Prototype 1

The first prototype design was created to have a snug fit against Eliza's torso. Shirts in her existing wardrobe had anywhere between 1 inch (2.54 cm) to 1¾ inches (4.45 cm) ease around the waist. Eliza demonstrated this made her feel uncomfortable by pulling at them, picking at their seams, and trying to take them off. After trying on Prototype 1, Eliza didn't show any signs of discomfort because it was tighter in fit. When the shirt was stretched away from her waist, the gap between the material and her body was ¾ inch (1.91 cm). The team tested shirt designs for comfort by comparing Prototype 1 with one of Eliza's existing shirts during a fitting. Members of the team noticed Eliza didn't want to change back into the looser shirt she arrived in, which reconfirmed the hypothesis that she preferred tighter-fitting clothing.

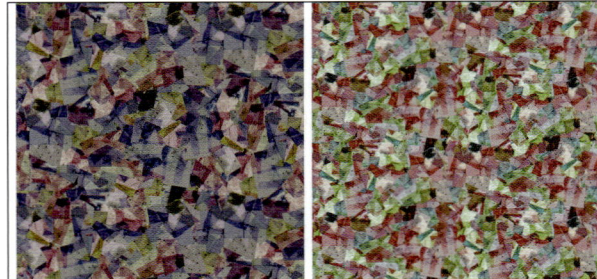

Figure 4.83 Eliza's mosaic artwork translated to custom fabric patterns. Courtesy of Open Style Lab.

wanted to celebrate Eliza's talents. Her collaging activities already demonstrated her artistic skills, which could translate into personalized products like the shirt design. The mosaic artwork Eliza created with her home therapist was digitally printed in a repetitive pattern onto the fabric to add a personal touch.

Prototype 3
Material Studies

The type of fabric chosen and the construction technique applied strongly determine the durability of the shirt design. To eliminate key areas that Eliza typically ripped apart, three options for seams were evaluated for their strength, feasibility of construction, and comfort: ultrasonic welding, silicone coating, and Bemis® bonded seams. Results showed that (1) ultrasonic welding created weak points within and immediately lateral to the seam, wearing down the integrity of the fibers, (2) Eliza enjoyed picking the silicone coating off the stitches (Tomchek and Dunn 2007), and (3) the Bemis® bonded seam could withstand tensile stress and strain significantly greater than Eliza's picking strength. Bemis® bonded raw edges also proved more effective for the neckline and sleeve hems due to their smoothness and low-profile texture.

Figure 4.82 Eliza testing different iterations of Prototype 1 for fit. Courtesy of Open Style Lab.

Prototype 2

Personalization can be achieved in two different ways: customizing the design and tailoring it to the body. Customization enables people to co-create a garment design. Members of the team recognized the value of personalization and

Prototypes 1 and 2 were constructed with Polartec® Power Dry® fabric, but the final series

Figure 4.84a–b Raglan sleeve and hem heat bonded with Bemis® to create bonded raw edges. Chrissy Glover places a bond strip on gray Polartec® fabric. Courtesy of Open Style Lab.

of shirts feature a combination of Polartec® Power Grid™, spandex jersey, and moisture-wicking compression jersey to give Eliza more color and style options. The fabrics used for the shirt design were light gray, dark gray, and a custom-printed pattern. As a final detail, the team added one more personalized design—the letter "E" for Eliza's initials in bonded material in replacement of an irritating clothing tag.

Materials
- Polartec® Power Grid™
- Spandex jersey
- Moisture-wicking compression jersey
- Bemis® bonded seams
- Bemis® bonded raw edges

Figure 4.85 Members of Team Ease Chrissy Glover and Uma Desai wearing the other two shirt designs in the Ease series. Courtesy of Open Style Lab.

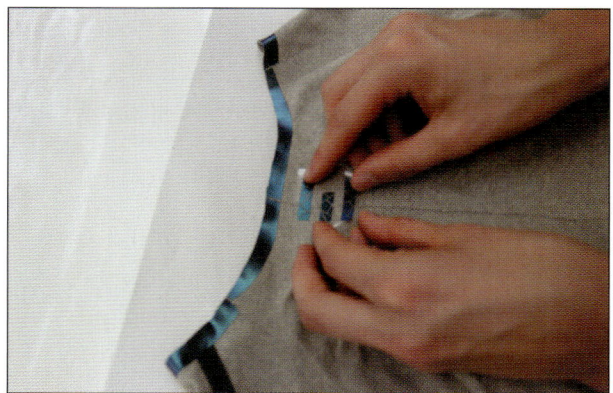

Figure 4.86a–b Chrissy Glover placing custom letter "E" design on two types of shirts. Courtesy of Open Style Lab.

Final Design Outcome

"It's the first shirt in two years that Eliza can't rip and take off."

Aimee Mury, Eliza's mother (pers. comm.)

Case Studies, Stories, and Interviews

Figure 4.87 a–b Aimee Mury communicating with Eliza Mury and Chrissy Glover on the final shirt design. Courtesy of Open Style Lab.

The team created two more copies of the same shirt design. The Ease shirts are made with various stretch fabrics in activewear-inspired shapes that accommodate Eliza's age, activity level, and personal style. They feature bonded seams and hem technology that she can't reach and pick at, as well as raglan sleeves with underarm gussets that allow for a greater range of motion. Finally, the garments are cut longer than normal to accommodate for growth and to give extra coverage if she pulls her shirt up.

References

Dunn, Winnie. 1999. *Sensory Profile*. Amsterdam: Elsevier.

Tomchek, Scott D., and Winnie Dunn. 2007. "Sensory Processing in Children with and without Autism: A Comparative Study Using the Short Sensory Profile." *American Journal of Occupational Therapy* 61: 190–200. doi:10.5014/ajot.61.2.190 (accessed August 14, 2023).

Modiste
Emily Ladau, Nicholas Paganelli, Yangkyoon (Nate) Oh

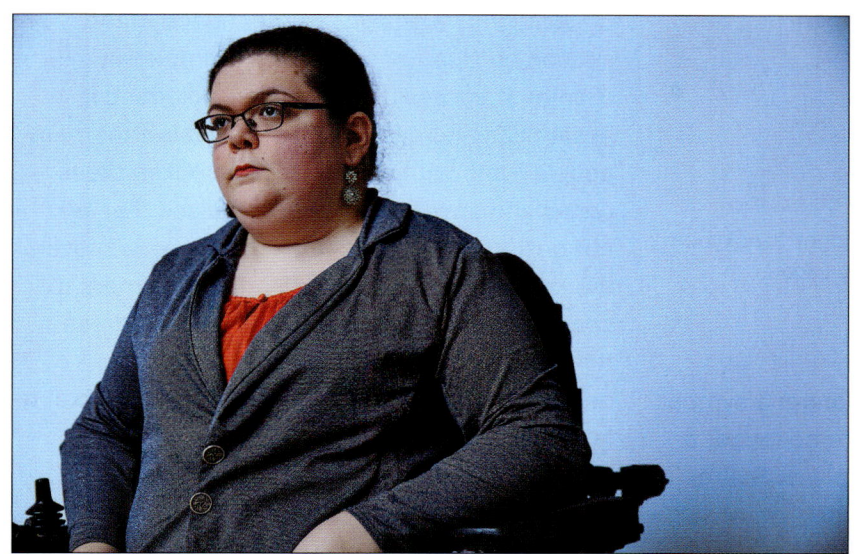 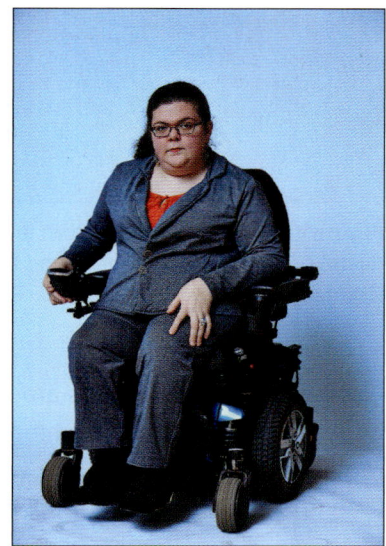

Figure 4.88a–b Emily Ladau wearing Modiste, a tailored suit, sitting against a light blue back drop. Photography by Kilian Son for Open Style Lab. Courtesy of Open Style Lab.

Key Functional Features
Moisture repellent
Antimicrobial
Discreetness

Summary
To create a tailored suit that provides comfort and allows for movement while maintaining a formal appearance.

Design Goal
To create a custom-fit jacket and pant suit that promotes greater mobility and addresses Emily's personal style.

Team Modiste
Emily Ladau: Client, subject-matter expert, and disability rights activist
Nicholas Paganelli: Design technologist and fashion designer
Yangkyoon (Nate) Oh: Fashion designer

Collaborator Acknowledgments
Parsons School of Design, Open Style Lab

Background
Fashion plays a significant role in the relationship between the person and the workplace. The power suit and pantsuits are a hallmark of American workwear. But in the world of men's tailoring—retailers, designers, and shoppers—the suit no longer represents power. "The power suit is dead," writes Robin Givhan for *The Washington*

Figure 4.89 Team Modiste from left to right: Yangkyoon (Nate) Oh, Nicholas Paganelli, and Emily Ladau. Courtesy of Open Style Lab.

Post, alluding to how suits are no longer a symbol of social position or rank as they are no longer a requirement in the corporate workplace. Rather than imbuing a position of power, they are about style and allude to political environments. For Emily, the suit designed with her team embodies both her personal style as a writer and disability rights activist. This collaboration examined the comfort, modesty, and fabric choices of bespoke suit-making and using fabrication tools to create adaptive professional clothing.

Introduction
Emily is an award-winning writer, speaker, and disability rights activist who received a BA in English from Adelphi University. She provides communications and social media strategy consulting as well as editorial services for multiple disability-related organizations and initiatives. Emily was born with Larsen syndrome, which often presents itself as a dislocation of large joints and skeletal malformations. Emily experiences muscle weakness and joint and bone contraction. Her elbows are also in a constant 90-degree position. Emily's choices in clothing must meet her bodily needs but also consider her electric wheelchair.

The Project
"I realized that this collaboration would be a learning experience and an opportunity to have something adapted to my exact needs," said Emily (Ladau, pers. comm.) about the importance of collaborative design. As mentioned in the section on Heartist and earlier examples in chapter 1, accessible professional clothing is essential to increasing **social inclusion** for many people living with disabilities. Unspoken dress codes and uniforms for the workplace that are not designed to consider disability needs prevent the completion of tasks and present barriers to social participation in work culture. To address this, the team started by first investigating personal biases and disability culture in relation to fashion design.

Design Process
1. Research and Discussions About Ableism
2. Goal Setting
3. Identifying Design Requirements and Challenges
4. Prototype Iterations

Research and Discussions About Ableism
A key factor in how clothing reflects people's status is designs that create a sense of uniformity, therefore accommodating a variety of body types in size, fit, and design. This is often seen in uniforms or suits that need custom tailoring. While tailored suits have long been in practice, proportions and fit remain frustrating challenges for PWD. Fit and comfort were two challenging aspects to design making, but also design perception. For example, during the first site visits, one of the learning points from Nate and Nicholas was Emily's donning process. The way Emily wore clothing was entirely comfortable despite it appearing uncomfortable from their perspective. Lack of representation of how clothing looks on disabled bodies creates an

Figure 4.90 Yangkyoon (Nate) Oh, Nicholas Paganelli, and Emily Ladau discussing suit styles and dressing routines. Courtesy of Grace Jun.

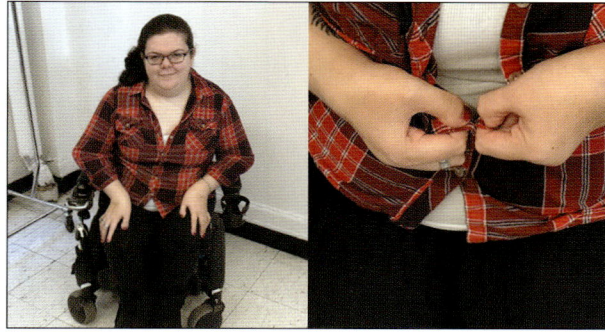

Figure 4.91 Emily Ladau in a red checked shirt that has small buttons. Courtesy of Nicholas Paganelli.

expensive and complicated guessing game to find clothes that fit well. Nicholas recalls an early meeting with Emily at her home and how she challenged his ideas of disability as inspiration, "Emily quickly helped us understand the subtleties of her experience and the simple solutions she had been using for years. This helped me appreciate that what she needed was not radically different clothing solutions, but just simple, nuanced ideas that took her experience into account" (Paganelli, pers. comm.). After several visits and sessions of trying on clothes, Nate and Nicholas began translating Emily's dressing behaviors into design sketches to begin prototyping.

Goal Setting
The team observed how Emily comfortably dons jackets, tops, and blazers. Her process involves sliding a single sleeve onto one forearm and holding the other by placing a hand into the armhole. She then flips the garment over her head, pushes her other arm into the sleeve, and twists her body to make it rest on her back before finally untwisting both sleeves. In addition, any garments with small buttons took more time to unbutton when donning.

Every visit revealed something new to Nate and Nicholas, like the fact that Emily was constantly buying and returning items because of materials that didn't fit well. So, they began to look outside her closet and home for more information. After paying close attention to the in-store shopping experience, they better understood the needs, challenges, and barriers faced by PWD. For instance, merchandise on high or low shelves couldn't be reached, aisles were often too narrow and/or cluttered to accommodate a wheelchair, and dressing rooms were almost always too small. In addition, the clothes being sold were typically difficult to don and doff due to the likes of small buttons and zipper heads, tight neck and armholes, and stiff fabrics.

Design Goals
- Sleek: using high-performance stretch material that looks fashionable
- Breathable: using material that can function well during long work events
- Comfortable: removing buttons and closures that make the garments difficult to wear

Prototype Iteration 1
To better understand the impact of speed, the team recorded how long it took for Emily to put on her jacket: twenty seconds on average, but only ten seconds for heavier winter coats with a

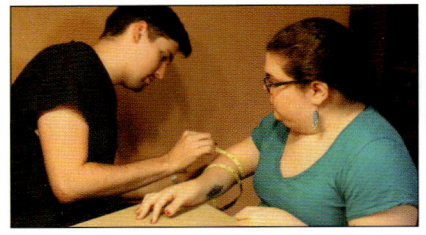  

Figure 4.92a–c Nicholas Paganelli taking measurements of Emily's forearm when creating a custom pattern design and using pins on draped prototype gray suit jacket. Courtesy of Nicholas Paganelli.

smooth lining. With this information, they set out to create a suit jacket prototype that would be quick to put on and take off. To accomplish this, they used several time-testing methods with different materials and features that informed them of two major things: smaller buttons slowed down the dressing process, while fabrics with stretch and/or a smooth surface sped it up.

Prototype Iteration 2

It was necessary to use a stretchy fabric with some synthetic content to provide a comfortable suit for Emily that could be donned quickly. Spandex, a fiber commonly found in athleisure wear and denim jeans, has increasingly been used in suits. Knit technologies and new fiber sciences provide ways that stretch can support different parts of the body. For instance, high-performance spandex has more stretch. A sleeve with this type of fabric could easily slide over Emily's elbows in a way that rigid fabrics cannot. Before testing their material choices, members of the team measured Emily's body. In doing so, they noticed a few additional measurements that might not appear on a standard measurement sheet. For example, Emily's arm was measured for overall length but also in segments so that her individual proportions could be respected.

Prototype Iteration 3

The team created several alternative possibilities using computerized knitting and fabric swatch making to create a custom suit more efficiently. Using computerized knitting provides an automated process for numerous garments because it does not require a human operator to pass the yarn back and forth. Garments can be designed, tailored or sized, and modeled in specific areas using computer software and 3D animations. These kinds of features are used in equipment produced by companies like Shima Seiki and Stoll to bring a garment from ideation to completion with mostly digital operation.

During the final fabric study, the team discovered that 70 percent cotton, 24 percent polyester, and 6 percent spandex (more than the average percentage of stretch found in Emily's closet) was the ideal combination for making a comfortably fitted suit. Additional integration of spandex would contribute to a faster dressing speed for Emily. Finally, the team documented Emily's donning and doffing time. The final design of the suit jacket took 48 percent less donning time and 50 percent less doffing time compared to her existing cardigan. This made it apparent that the process of donning and doffing is directly impacted by what kind of fabric is used. Polyester—a smooth synthetic fiber—provided a sleek look and feel despite being a stretch knit.

Figure 4.93 Computerizing knitting swatches created by Nicholas Paganelli using the Shima Seiki. Courtesy of Nicholas Paganelli.

Materials
- Cotton
- Polyester
- Spandex
- Knitting machine
- Knitting computer software

Final Design Outcome

"How I felt putting the jacket on, I knew that it was going to be an exciting process."

Emily Ladau (pers. comm.)

Modiste is a standard suit pattern with modifications to the pant leg length, hip width, and rise in the back (to provide additional coverage when in a seated position). The design of the jacket was engineered solely to the circumference of the armhole, which was a significant factor in donning and doffing for Emily based on research and observation of her existing wardrobe. The armhole was the primary source of difficulty in the fitting process, as the quality of its fit and balance affected many other components of the jacket.

Revolve
April Coughlin, Kaitlin Crowther, Renata Gaui, Maggie Mahoney

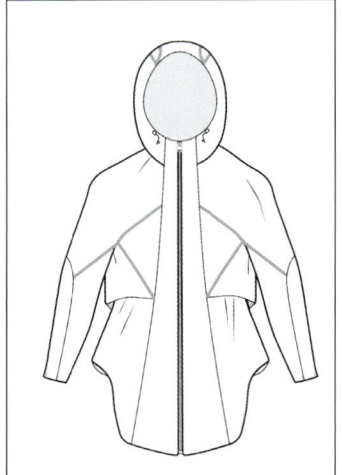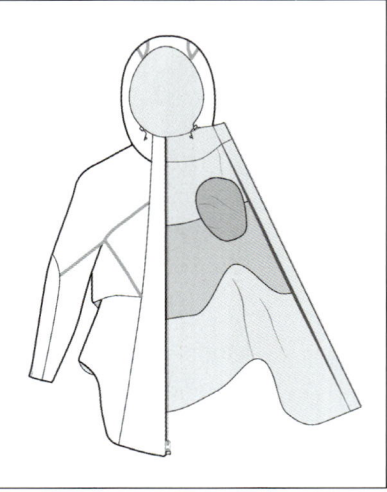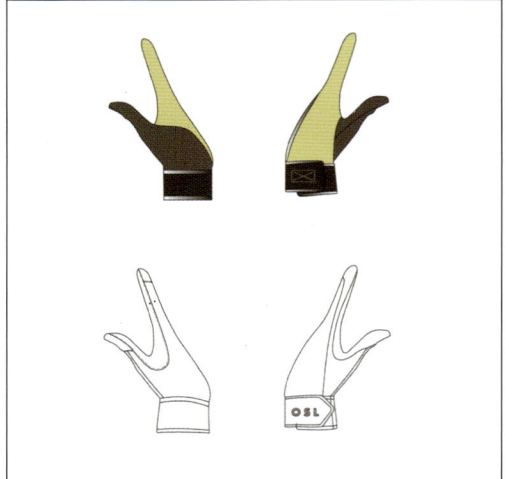

Figure 4.94a–c Vector illustrations of Revolve's jacket and glove covers drawn by Maggie Mahoney. Courtesy of Open Style Lab and Maggie Mahoney.

Key Functional Features
Durability
Water-repellent
Attention signaling

Summary
How to create an inclusive garment for wheelchair users that's warm, weather-resistant, and withstands everyday wear and tear.

Design Goal
To create a protective outdoor garment that meets the demands of even the most extreme wheelchair users during their city commutes; this includes increasing visibility, making it waterproof, and enhancing durability.

Team Revolve
April Coughlin: Client, subject-matter expert, educator, and disability rights activist
Kaitlin Crowther: Occupational therapist
Renata Gaui: Designer and engineer
Maggie Mahoney: Fashion designer

Collaborator Acknowledgments
Parsons School of Design, Open Style Lab, Polartec®

Background
Wheelchair users who regularly commute to and through a city are no strangers to various weather conditions and long distances due to a lack of accessibility points throughout public transportation. As a result, PWD may endure danger, discomfort, and anxiousness when they travel back and forth. Current solutions available in the market are either unsatisfactory, ineffective, or unappealing.

For this design collaboration, the research focuses on the specific needs of April Coughlin, a woman who has used a manual wheelchair for more than 30 years. Through a collaborative methodology and deep dive into April's lived experiences, Team Revolve identified the following design opportunities: (1) increase her visibility during commutes to ensure safety, (2) keep her dry throughout every season in the northeast (The American Occupational Therapy Association 2020), and (3) extend the lifespan of her jacket and gloves. To achieve these goals, the team researched and developed wearable designs that evolved into a waterproof raincoat and glove guards.

Figure 4.95 Team Revolve members from left to right: April Coughlin, Kaitlin Crowther, Renata Gaui, and Maggie Mahoney. Courtesy of Open Style Lab.

Introduction

April is busy as an activist for disability rights, as well as a professor who teaches courses on disability studies at Long Island University and SUNY New Paltz in New York. While spinal cord injuries affect everyone differently due to their location (Pendleton and Schultz-Krohn 2013), her SCI is at T10 of her thoracic spine, which affects the function of her lower extremities. To get from place to place, her daily life consists of using a customized manual wheelchair with the strength of her upper body. She propels and brakes her wheels with her hands and forearms, but eventually wears down her gloves and jacket sleeves from the repetitive abrasion.

April lives with her husband in an apartment that has been augmented to be wheelchair accessible; this includes lower sinks, several grab bars, and a shower bench. She is independent in all ADLs and IADLs, which she had seemed to master with creative systems and techniques. To do this, she found and tested a variety of off-the-shelf products that help her improve her existing routines. For example, she attaches three different bags to her wheelchair, each with a designated purpose. Her unique compartment system helps her quickly and easily access what she needs, when she needs it throughout the day.

The item she approached Open Style Lab for help with was a raincoat that covers her body while accommodating how she uses her wheelchair. She explained, "Unfortunately, rain jackets aren't made for wheelers because we're sitting down, and they don't cover our laps. So, while our top half may stay dry, our lower half gets completely soaked." Her solution has been to wear a standard rain jacket (that supports her mobility) with a trash bag over her lap (to protect her legs and feet). Ultimately, this method is ineffective and demeaning.

Design Process
1. Research
2. Observation and Interview
3. Goal Setting
4. Journey Map
5. Identifying Design Requirements and Challenges
6. Prototype Planning and Development

Case Studies, Stories, and Interviews

Design Requirements
1. Protects the user from precipitation and wind
2. Specific fit for wheelchair users:
   a. Sits comfortably on a seated body
   b. Doesn't feature loose fabric that could get tangled in the moving chair
   c. Does feature plenty of room in the arms and shoulders for propelling the chair
3. Durability: stands up to abrasions that cause ready-made garments to tear and wear out
4. Visibility: provides extra safety for the user in high-traffic areas

Part 1: The Raincoat

Weather Protection

April's initial primary concern was about staying dry when she commuted to and through New York. "As a wheeler, I can't push my chair and carry an umbrella at the same time," explained April (Coughlin, pers. comm.). Her trash bag solution didn't work because the rainwater always pooled on her lap and leaked onto her seat cushion, which made her feel uncomfortable. For some individuals with an SCI, this is a particular concern because they're at higher risk for developing a pressure ulcer—the most common complication secondary to a spinal cord injury (NSCISC 2006).

When other team members first learned of April's dilemma, they thought it was a great opportunity to problem-solve with design. To create a product that would address her specific needs, the team members first needed to find a material that was waterproof, functional, and comfortable to wear while commuting. After researching viable and innovative options for April's raincoat, they all agreed to use Polartec® Power Shield® Pro—a double-sided material that's hydrophobic on one side and breathable on the other. Additionally, the raincoat's construction had to address the flow of water by integrating a means of funneling it away from April, her wheelchair, and her belongings—even when donning and doffing the garment.

Fit

One of the raincoat's most important design elements was how it fit April. From the very beginning, Team Revolve had to address a common challenge for wheelchair users: loose material getting stuck in the wheels. Because this could cause her wheelchair to tip, flip, and result in physical harm, they had to make sure her garment was snug—especially around the torso. In direct contrast, however, it also needed room to let her shoulders, elbows, and wrists help propel, increase, and decrease the speed of her manual wheelchair; otherwise, her upper body movements would tear the seams. So, the team ultimately had to balance the importance of the raincoat's slim fit and its range of motion (see fig. 4.96).

Durability

Wheelchair users who commute through any city are often exposed to extreme weather conditions for extended periods of time, in part because of an inaccessible subway system that doesn't feature many stations with a working elevator. When wheelchair users are in transit through the streets, the last thing they should have to worry about is that their clothing might fail them or tear. While designing April's raincoat, the team knew it

Figure 4.96a–b Raincoat prototype iterations in muslin fabric. Courtesy of Maggie Mahoney.

Figure 4.97 April Coughlin wearing the Revolve raincoat with abrasion-resistant fabric.
Courtesy of Open Style Lab.

was essential to consider the repetition of her movements and how it affected the longevity of her clothes. Since her coats tended to tear underneath her arms, they incorporated a stretchy fabric, created extra space around the shoulders, and inserted lambskin leather (a highly durable material) onto the underside of the forearms to resist abrasion from April's wheels (see fig. 4.97).

Visibility
Another important design element was the raincoat's visibility—a major safety concern for many wheelchair users. In a 2015 study conducted at Georgetown University, it was discovered that wheelchair users are a third more likely to be killed crossing the crosswalk (Poon 2015). To increase the visibility of April's raincoat, the team created a reflective seam tape accent (see fig. 4.98b).

Prototype Iterations
Team Revolve used the patterns and measurements of April's favorite jackets and coats to create the first iteration of her raincoat. In addition, fabric was draped over the lap and back of the chair that didn't catch in the wheels, different types of hoods were explored, and several fittings and wear tests informed the final prototype.

Material Studies
Throughout the design process, Team Revolve was introduced to several innovative materials and manufacturers. After thorough research and waterproof testing, they decided to use Polartec® Power Shield® Pro plus a more durable version to reinforce the elbows and forearms. Additional rounds of user testing following the construction of the final prototype showed that the raincoat became very heavy when wet, so a lighter material with the same breathability was used in later designs.

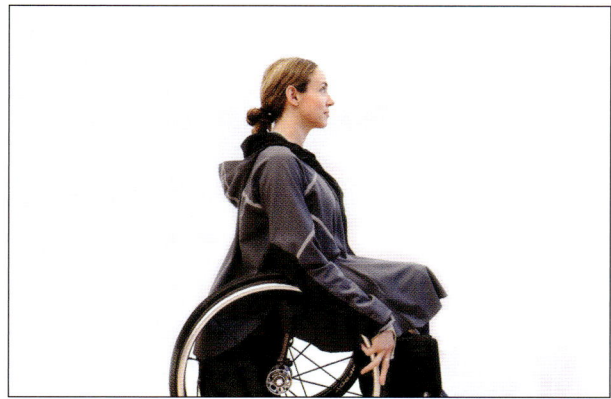

Figure 4.98 a–b Reflective seam tape from Bemis® Bonds (courtesy of Renata Gaui) applied to Revolve raincoat for high visibility. Courtesy of Open Style Lab.

Case Studies, Stories, and Interviews

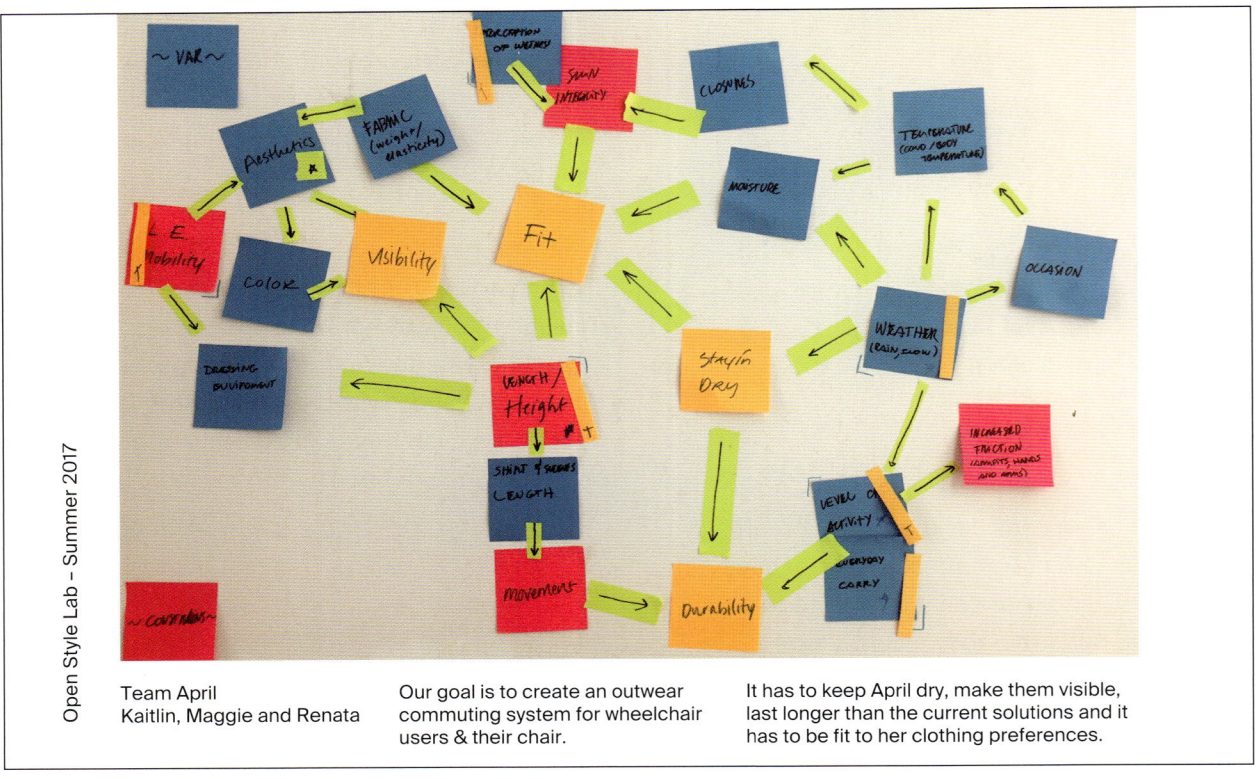

Figure 4.99 Team Revolve researching various performance materials based on design requirement goals for visibility, fit, and durability. Courtesy of Open Style Lab.

Final Design Outcome

April's Revolve raincoat is made with Polartec® Power Shield® Pro—a durable double-sided material that's hydrophobic on one side and breathable on the other. Adjustable channel systems are incorporated throughout the garment to prevent water from pooling in the lap and the wearer from getting wet when donning and doffing. It's structured to provide room for shoulder and arm movement, reinforced with durable fabric to resist abrasion, measured to form a slim silhouette that doesn't get caught in the wheels, and features reflective elements to ensure visibility and safety.

Part 2: The Glove Guard

Durability

The second garment issue April had was that her gloves would fall apart. "The gloves I buy just can't seem to keep up with me. They rip over and over in the same areas, and I need them to handle what I put them through," explained April (Coughlin, pers. comm.). She consistently wore them when she left her apartment to protect her hands from the friction of her wheels, as well as shield them from dirt picked up in the streets. April told her team that she would wear down a new pair of gloves in two weeks, so they got to work researching what was already available in the market (see fig. 4.100).

After examining golfing gloves, batting gloves, rubber gloves, Kevlar (protective material known for its cut-resistant properties) gloves, and

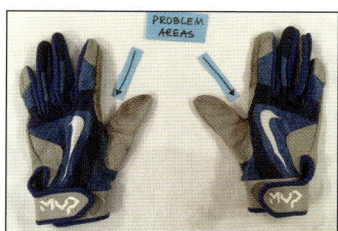

Figure 4.100 A photo of April's gloves by Team Revolve to examine precise areas in the hand that needed extra durability. Courtesy of April Coughlin.

Figure 4.101 Various glove cover designs created with machine knitting in Kevlar and cotton thread. Courtesy of Renata Gaui.

nitrile-coated gloves, they discovered a few common design flaws. One was that those with patterns at the seams, particularly along the index finger and thumb, contributed to the item's breakdown. From such learnings, the team then opted to create gloves that would guard April's existing gloves rather than replacing them altogether. Aided by skilled technicians at Parsons School of Design, Team Revolve made knitted gloves without seams along the index finger and thumb that were highly durable and protective of her current gloves.

Prototype Iterations
Team Revolve was able to identify that April's existing gloves mostly ripped along the inseam between her index fingers and thumbs. They tried to coat them with an abrasion-resistant material, but it proved to be too sticky and impeded her wheelchair's mobility. So, they worked with technicians at Parsons School of Design to develop a glove cover without seams along the index finger and thumb that was made from a custom mixture of Kevlar and cotton thread.

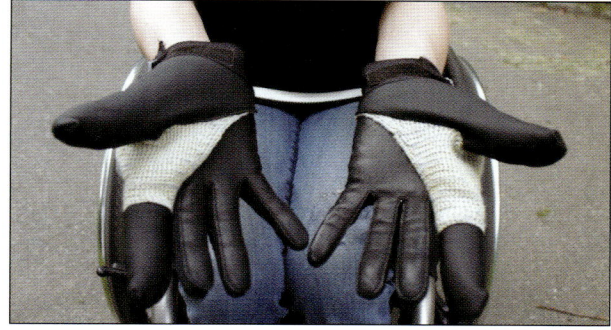

Figure 4.102 Final glove designs created in Kevlar and abrasion-resistant materials. Courtesy of Renata Gaui.

Material Studies
The team extensively researched machine knitting and industrial-strength coatings. After purchasing and testing (with April) almost every kind of kitchen and work glove for sale, they determined that knife-proof wasn't good enough. So, they experimented with Kevlar threads and abrasion-resistant materials to determine the best design options.

Project Materials
- Polartec® Power Shield® Pro
- High-abrasion tent-repair fabric (elbow section)
- Pull tabs
- Zippers
- Kevlar
- Cotton
- Reflective tape
- Elastic tape
- Velcro®
- Leather
- Neoprene

 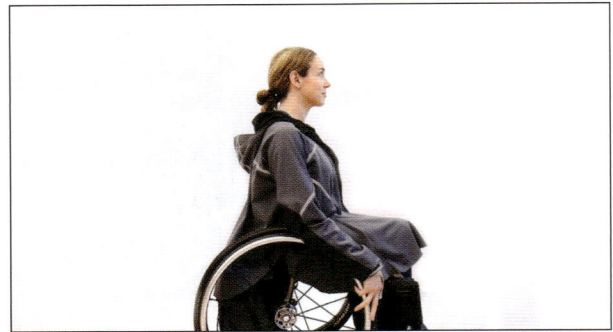

Figure 4.103a–b April Coughlin sitting in her manual wheelchair, wearing final glove designs and Revolve raincoat. Courtesy of Open Style Lab.

Final Design Outcome

Team Revolve created a knitted glove cover without seams along the index finger and thumb. This small detail ultimately reduced chafing, irritation, and seam breakage while increasing the durability and comfort of April's gloves during her commute.

Project Outcome

Creating functional clothing for commuters in wheelchairs is something the fashion industry has failed to do. Open Style Lab formed Team Revolve, made up of one fashion designer, one engineer, one occupational therapist, and one client, that succeeded in creating two innovative products that addressed the specific needs of wheelchair users. The raincoat was designed to comfortably protect the wearer from their head to their knees, and the glove guards were designed to increase the lifespan of their gloves.

References

NSCISC (National Spinal Cord Injury Statistical Center). 2006. *2006 NSCISC Annual Report for the Model Spinal Cord Injury Care Systems*. Birmingham, AL: National Spinal Cord Injury Statistical Center. https://www.nscisc.uab.edu/PublicDocuments/reports/pdf/NSCIC%20Annual%2006.pdf (accessed July 29, 2017).

Pendleton, Heidi McHugh, and Winifred Schultz-Krohn. 2013. *Pedretti's Occupational Therapy: Practice Skills for Physical Dysfunction*, 7th ed. St Louis, MO: Elsevier.

Poon, Linda. 2015. "Why Are Wheelchair Users More Likely to Be Killed in Traffic Than Other Pedestrians?" *Bloomberg*, November 19, 2015. https://www.bloomberg.com/news/articles/2015-11-19/study-pedestrians-in-wheelchairs-are-a-third-more-likely-to-be-killed-in-road-accidents (accessed August 08, 2017).

The American Occupational Therapy Association. 2020. "Occupational Therapy Practice Framework: Domain and Process Fourth Edition." *American Journal of Occupational Therapy* 74 (2): 1–87. doi:10.5014/ajot.2020.74S2001 (accessed August 14, 2023).

Interview
Interactive Garments and Textiles— Dr. Jeanne Tan

Figure 4.104 Dr. Jeanne Tan, Associate Professor of School of Fashion and Textiles providing a lecture. Courtesy of Dr. Jeanne Tan.

Dr. Jeanne Tan possesses dual roles, firstly as Chief Operating Officer and Centre Assistant Director of the Laboratory for Artificial Intelligence in Design (AiDLab) and Associate Professor of School of Fashion and Textiles, the Hong Kong Polytechnic University. Jeanne's work investigates the interface of design and technology. She often utilizes textiles and fashion as a communicative platform, integrating traditional craft and engineering as the syntax of the creation's narrative. Her research interests include **smart material** design, interactive textiles for well-being, hybrid design processes, and smart wearables.

Grace
What is your role, and can you explain what you are currently working on?

Jeanne
My name is Jeanne Tan, and I hold dual roles at the university. My first role is Chief Operating Officer at the AiDLab. The laboratory is funded by the InnoHK research clusters, located in Hong Kong Science Park. Hong Kong Science Park is the innovation and technology hub of Hong Kong. InnoHK is a major initiative of the Hong Kong Special Administrative Region Government to develop Hong Kong as the hub for global research collaboration. My second role is as Associate Professor at the School of Fashion and Textiles at the HKPOLYU (Hong Kong Polytechnic University).

I have a PhD in textiles from a very traditional art school, Glasgow School of Art. I started integrating technology into my work when I started my academic career at the university. There were many opportunities for collaboration with colleagues of different disciplines. I am a designer who uses AI (artificial intelligence) to transform conventionally passive textiles into interactive platforms.

Grace
What kind of research have you been recently working on?

Jeanne
The research I've been working on focuses on the integration of AI into textiles. My team and I have come up with the first contactless gesture recognition textile system that gives immediate illumination feedback. Part of the system is an illuminating textile knitted from POFs (polymeric optical fiber) and textile-based yarns. We have two main areas of patents for this invention: a gesture recognition system and the POF knitted structure. Instead of weaving the textile, which is the conventional way of making POF textiles, we are knitting them. Knitting enables us to create an illuminating textile with stretch, soft handle, and a more sustainable means of production as I only use the materials that I need to fashion into the final products. There is no waste from the cut and sew process. Another key innovation is that we can produce the textile on industrial flatbed knitting machines, which enables mass production and mass adoption. We have also developed techniques in which we can knit the illuminating textiles in a wide variety of weights, textures, and patterns.

Grace
The advancements in technology that you are researching in fashion and textiles are incredible. Do you think that this could also be applicable to people living with disabilities or the aging population?

Jeanne
Yes. We have licensed our technology to the Hong Kong Sheng Kung Hui Welfare Council for installation into their Wong Tai Sin District Healthcare Centre. They have installed a sensory wall made from our intelligent textile system as an interactive tool for sensory stimulation. Our textiles are soft to handle and could easily integrate into everyday products. We can program our system to recognize a wide variety of hand and body gestures, providing a very intuitive means for interaction. There is no need for complex textual instructions, and users of all ages, backgrounds, and abilities will be able to engage with it.

Grace
Can you give an example of where this fabric can be applicable and used by people in the aging population?

Jeanne
The intelligent textile system can be applied for rehabilitation purposes, especially within the context of multi-sensory therapy, which is beneficial for users with cognitive challenges—for example, users with dementia and learning challenges.

Grace
How do you see this research and application benefiting future students? For example, students who would be interested in adaptive fashion or designing for social change.

Jeanne
I come from a fashion and textiles background, and traditionally, designers were trained to practice within their disciplines. With the advancement of technology, we see a shift in the way in which we create, where we can push the boundaries of design and technology. As we become "polycreatives," we can also harness the power of AI as a medium for arts and crafts. Students can hone their skills in communication to enable greater group synergies, and this will enable more interdisciplinary explorations for creative innovations.

Grace
Where do you think technology intersecting with textiles is going? How do you think it will evolve in the future based on your research?

Jeanne
I think it is a very exciting time for innovation right now. The pandemic changed the way in which we live, consume, and communicate. We want more time to invest in our well-being. Technology can be integrated into many aspects of our lives in terms of wearables, products, and services that can make the "process" of life more effortless. There is great potential for a common and omnipresent material like textiles to be transformed into a platform that adapts to the needs of the users. This will be a more sustainable way of consuming products that is capable of customization instead of constantly changing products as your needs evolve.

Grace
Adaptation is important in design. What does adaptive mean to you in the context of fashion?

Jeanne
Adaptive fashion means, to me, a way to adapt according to the needs of the user. It is something that can evolve with your needs over time.

Versa Vest
Douglas Balder, Pamela Cooper, Andrew James Sapala, Ying Xiao, Fanyun Peng

Figure 4.105 Team Versa Vest from left to right: Andrew James Sapala, Douglas Balder, Pamela Cooper, Ying Xiao, and Fanyun Peng. Courtesy of Douglas Balder.

Key Functional Features
Comfort
Stability support
Weight

Summary
A specialized torso wrap using a rubber bag apparatus and tailored foam.

Design Goal
To create a vest that's comfortable, conservative, washable, waterproof, and easy to don and doff.

Team Doug
Douglas Balder: Client and subject-matter expert, architect, disability rights activist
Pamela Cooper: Fashion designer
Andrew James Sapala: Sculpture designer
Ying Xiao: Design technologist
Fanyun Peng: Design technologist

Background
From weddings to sports, vests can be found at all types of events. Their basic design allows sufficient arm mobility while keeping the body's core covered. In cooler temperatures, they serve as a great mid-layer between a long-sleeved base and outer jacket. In addition, vests also facilitate spine support and upper body protection of a person's back. The Versa Vest team explored a vest design that provided comfort and adjustable back support while complying with Doug's conservative sense of style: a chic black vest to reduce the discomfort in his back when sitting for prolonged hours.

Introduction
Doug is an architect and community organizer. In 2015, he was diagnosed with multiple myeloma, a treatable but incurable blood cancer, which led to eight fractured and compressed vertebrae and a height reduction of 5½ inches (13.97 cm). Spinal cord compression and bone pain in his lower back have impacted his life by making it difficult and painful to sit or stand for long periods of time. It has also caused a physical change to Doug's back, resulting in an outward curvature of his spinal column, essentially compressing in on itself. His constant pain restricts his range of motion and limits his daily activities, so he wanted to create a wearable solution that reduced his discomfort. With his partner Joan, Doug investigated making padded pillow-like cushions that could be inserted into a custom vest.

Doug's sleek city style includes versatile garments in cool colors such as navy, gray, and

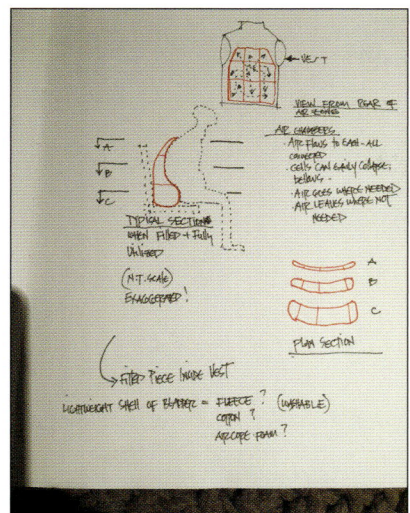 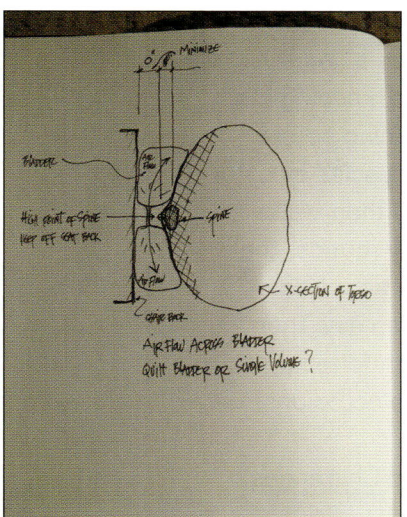 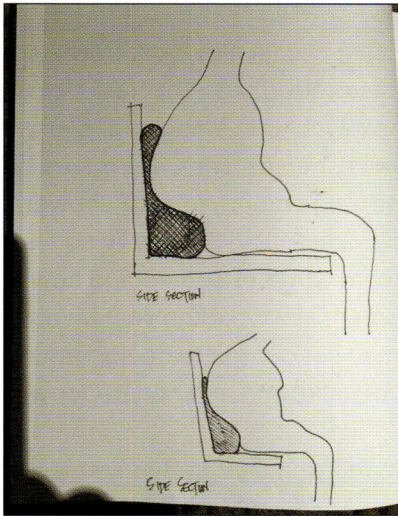

Figure 4.106a–c Sketches of designs that could provide spinal support for seated bodies drawn by Douglas Balder. Courtesy of Douglas Balder.

midnight black. Simple dark and textured shades were also conveyed in the design process with the Versa Vest team.

Design Process
1. Research
2. Observation and Interview
3. Goal Setting
4. Journey Map
5. Identifying Design Requirements and Challenges
6. Prototype Planning and Development

Research
The team began with observations, discussions, and an overall study of Doug's existing wardrobe. His clothing style consisted of cotton and wool sweaters, waterproof outer coats, several vests, and simple undershirts, primarily in white, black, and neutral colors. While Doug's existing vest solution from Uniqlo was a promising starting point for design, it had a long hem and was too bulky in the back. The zippers were also too small, making it difficult to maneuver in cold weather. Studies based on this vest led the team to research garments that don't constrict the stomach or lower torso area while sitting down, yet still offer the back support Doug needed. In addition, they examined protective gear that used air bags and foam inserts for support, such as Nudown and Titin, that could better support Doug's curved spine.

Observation and Interview
Multiple myeloma causes Doug's spine to become increasingly fragile, so strenuous activities contribute to his pain and fatigue. The team met with him to better understand his specific needs, daily routine, and personal style—noting his preference for activewear with cool colors and conservative silhouettes.

Goal Setting
The short-term goal was to create an adjustable vest with back support in order to relieve Doug's pain when he sits, thus enabling him to comfortably work long days at a desk. The long-term goal was to increase his range of motion with daily use of the back vest.

Case Studies, Stories, and Interviews   **169**

Journey Map

Doug chooses to work from home on a customized desk that lets him stand, keeping his spinal column elongated. Every morning before work, he stretches, warms up his muscles by moving around, takes a shower, and then walks his dog, Ruby. By observing his daily routine, the team came to understand how long he would potentially wear the vest on any given day. Then, they set out to create something comfortable, versatile, and appropriate for various occasions.

Identifying Design Requirements and Challenges
1. Versatile for year-round use.
2. Easy to don and doff.
3. Durable enough to withstand abrasions and support the wearer's curved spine.
4. Comfortable without putting pressure on the wearer's back.

Prototype Iteration 1

The team's first prototype used a combination of lighter and sturdier kinds of foam, ranging from mattress material to EVA (ethylene-vinyl acetate) copolymer foam to provide comfort and support. After it was primed and sealed, it was painted black and fitted into a jersey vest created with Doug's precise measurements. The back "spine-like" cushion allows him to lean forward and back while seated, giving him support in the section of his back closer to his neck. While the goal was to create foam placement that would isolate Doug's painful areas and distribute force along its flatter pieces, the prototype only relieved the discomfort in his mid-section—not his lower back. Additionally, the foam pieces were too bulky and didn't align with Doug's personal aesthetics.

Prototype Iteration 2

The team's second prototype leveraged each person's unique background in design to create a garment that blended art, sculpture, fashion, and technology. First, they used a skin-safe alginate to make a plaster cast of Doug's back, focusing on his most sensitive areas. The material, called Alja-Safe™, helped create a plaster negative, which the team could reference for design prototyping without meeting Doug at his studio. The plaster negative also proved useful when the team needed to augment the scale or pattern of the foam or textiles when prototyping. This allowed for two essential steps in the design process: (1) further developments when Doug could not be present and (2) a resin shell (The American Occupational Therapy Association 2020), which they ultimately discovered was uncomfortable because it provided very little stretch.

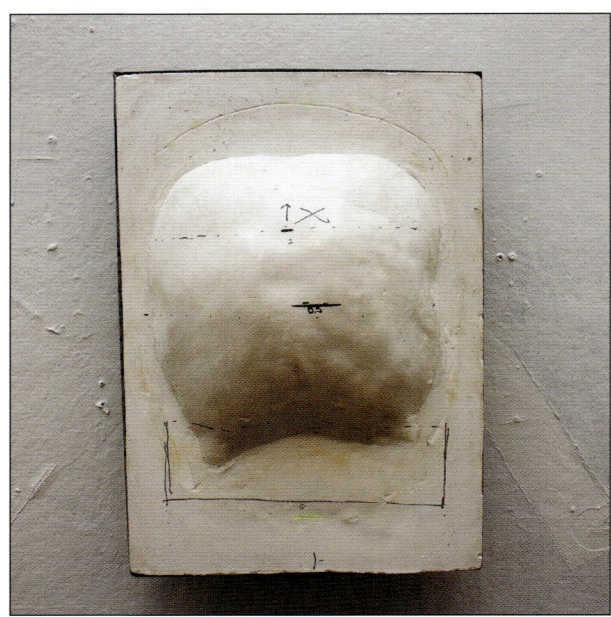

Figure 4.107 A cast of Doug's back designed by Andrew James Sapala. Courtesy of Andrew James Sapala.

New Tools

An airbag was placed in an envelope and tested with different amounts of synthetic foam. This proved to be a difficult situation, and the team quickly realized that they needed a soft, squishy material that could withstand impact without

Figure 4.108 Foam hidden inside a khaki-colored encasing and airbag enclosed inside the black encasings. A black pump designed to provide air into the encases created by Pamela Cooper and Andrew James Sapala. Courtesy of Pamela Cooper.

looking voluminous. While the EVA foam was too rigid, they found a latex foam that worked well, though it was extremely hard to cut in a precise manner. After they successfully cut the foam and encapsulated it with a washable material, they coated it with a soft, sheer material. Then, a three-dimensional wooden guide was built to push an upholstery through to keep the foam hidden inside the garment.

The team also purchased a seat cushion to deconstruct. This allowed them to cut open an airbag, which was made from durable rubber that could withstand high pressure and stay filled for a long period of time without leaking, and tailor it according to the dimensions of Doug's lower back. This airbag was then filled with latex foam and ultimately covered with a layer of tan neoprene. While the result was still bulky and unsightly, it proved to be the most comfortable solution.

Prototype Iteration 3

During the development of the third prototype, the team consulted with physical therapists from Open Style Lab and NYU Steinhardt's Department of Physical Therapy. They provided feedback about Doug's range of motion and suggested designing a wrap similar to what people use to bandage a sprained ankle or arm. Through this process, Pamela created a thin wrap design using stretchy jersey material. The front two wings of the garment could be attached with safety pins and pulled as tightly as Doug wanted, bringing the foam airbag closer to his lower back. While it was the most successful prototype, the foam airbag still moved too much when he transitioned from sitting to standing. The team also learned the jersey wrap material was too thin for Doug.

Prototype Iteration 4

After testing the wrap garment with its engineered airbag and foam, the team began constructing their fourth prototype. They used compression neoprene to design a wrap-vest hybrid. Then, a right wing (installed with a magnet for easy donning and doffing) was attached and pulled through an opening on the lower left side. The airbag and foam piece were placed into a pouch (made from neoprene and held together with magnets) inside the back of the garment. From a very small hole punched in the left side, the airbag's rubber tubing and pump were fed through for easy access. From there, the pump could be holstered in the sleeve of the left wing, as seen in figures 4.109a–b.

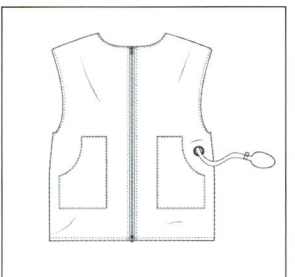 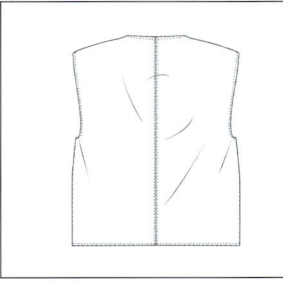

Figure 4.109a–b Versa Vest Illustration with an airbag's rubber tubing and pump. Courtesy of Pamela Cooper.

Material Exploration and Pressure Studies
Ying and Fanyun created a grid on a piece of muslin with embedded pressure sensors that connected to an Arduino and attached to Doug's lower back. By lightly pushing on different coordinates, the team was able to map out and isolate his pain points using a scale from 0 (extremely sensitive) to 30 (mildly sensitive).

They quickly discovered that they needed to implement a combination of foam and air around the extremely sensitive areas for the final wearable solution design. While the foam piece would always stay put, an airbag would be inflated or deflated as needed by an attached pump (for example, when Doug sits against a flat surface). Finally, the last design iteration incorporated a soft, durable material called neoprene.

Project Materials
- Alja-Safe™
- Neoprene
- Jersey and cotton fabric
- Nylon spandex fabric
- Rubber
- Latex and memory foam
- Fiber fill
- Down feathers
- Bluetooth® microchip
- Soldering paste and wires
- Arduino IDE program
- CO_2 cartridges
- Battery pack

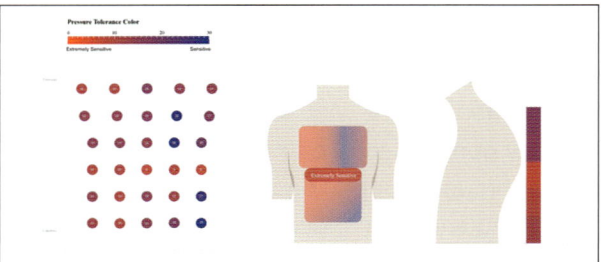

Figure 4.110 Visual sketch of isolated pain points on Douglas's back created by Ying and Fanyun. The chart and colors indicate "pressure tolerance".
Courtesy of Ying Xiao and Fanyun Peng.

Final Design Outcome
The Versa Vest features soft, waterproof neoprene on the exterior part of the garment. Its gray and black color complements Doug's personal dress aesthetic and provides protection against stains. The airbag and foam inserts were attributes that allowed Doug to be comfortable and supported while on the move. He was ultimately able to sit down for longer periods of time because the vest relieved the pressure he felt on his lower back.

References
The American Occupational Therapy Association. 2020. "Occupational Therapy Practice Framework: Domain and Process Fourth Edition." *American Journal of Occupational Therapy* 74 (2): 1–87. doi:10.5014/ajot.2020.74S2001 (accessed August 14, 2023).

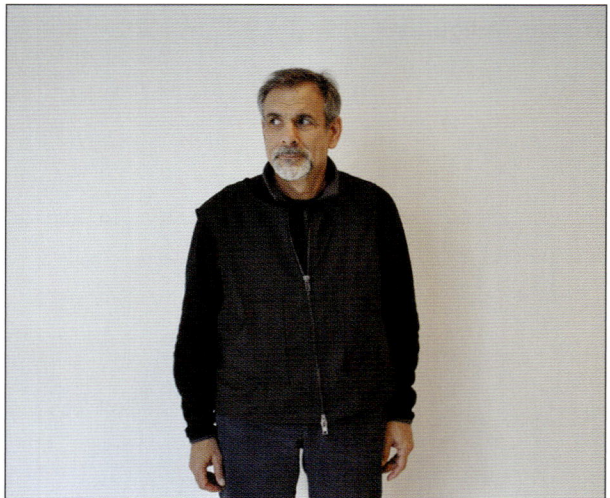

Figure 4.111 Douglas Balder wearing Versa Vest over a long-sleeve shirt with gray-blue pants.
Courtesy of Pamela Cooper.

Avisly
Mia Paget, Audrey Bosquet, Sam Cocjin, Arash Kani

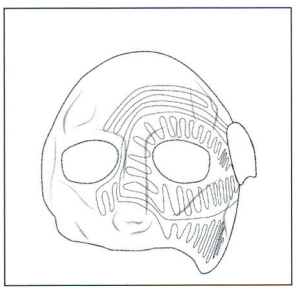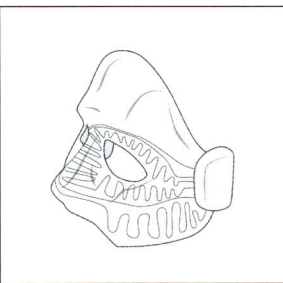

Figure 4.112a–b Vector illustration of Avisly mask drawn by Audrey Bosquet.
Courtesy of Mia Paget and Audrey Bosquet.

Key Functional Features
Connectivity
Protection
Heat
Comfort

Summary
Avisly is a transparent heated face mask for people experiencing chronic and/or acute pain in different areas of their face.

Design Goal
To create a face mask that prevents cold airflow to the face, blocks wind that causes pain sensitivity, and looks less conspicuous.

Team Avisly
Mia Paget: Client and subject-matter expert
Audrey Bosquet: Designer
Sam Cocjin: Occupational therapist
Arash Kani: Engineer

Collaborator Acknowledgments
Open Style Lab, Texas Instruments, Polartec®

Design Attributes
- Wind protection
- Heat conservation
- Style

Background
Protective face coverings were used more than 2,000 years ago in ancient Rome and before that as decorative accessories for spiritual rituals and performances in China. Like masks, face coverings have a protective function for individuals with specific occupations (e.g., doctors and construction workers), but since 2020, wearing them has become more normalized than ever. "It's a different world [after COVID-19] because now I won't get kicked out of a store for wearing a mask. I used to because I looked scary, and [people thought] I might be there to steal something," says Mia (Paget, pers. comm.). In this collaborative design project, the functional and aesthetic opportunities of mask design are explored by examining personal use and public perception.

Introduction
Mia is an engineer who experienced a traumatic injury while being treated in the emergency room for an ulcer. An NG (nasogastric tube) tube was poorly placed, breaking her middle turbinate and causing internal scarring and permanent nerve damage on the left side of her face. She then developed atypical facial neuropathy, or trigeminal neuralgia, without any medical resources (other than painkillers) to help her manage her chronic pain. Overuse of medications has caused side effects, ultimately making it difficult for Mia to focus long enough to complete tasks.

Case Studies, Stories, and Interviews

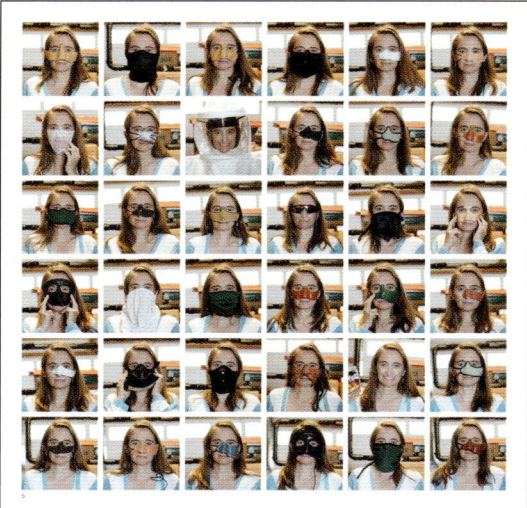 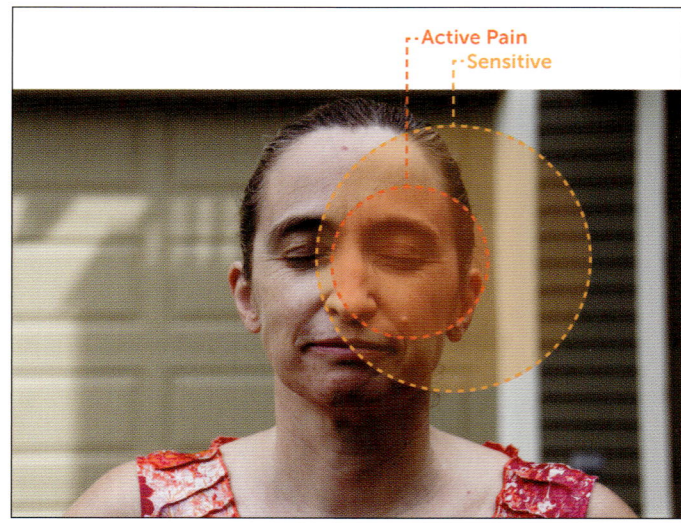

Figure 4.113a–b Various photographs of Mia Paget wearing masks she had designed to cover her face. Close-up photo of Mia's face detailing areas with active pain and/or sensitivity. Courtesy of Mia Paget and Audrey Bosquet.

Drawing from her engineering background, Mia created everyday solutions for herself, such as leather masks that help alleviate the pain from cold weather, harsh winds, and air-conditioned environments. While they serve their purpose, they're not as warm as she'd like them to be. So, she's explored ways to use heat to manage the nerve pain on the left side of her face, but the results have proved to be aesthetically challenging and unreliable. In addition to not addressing all of Mia's needs, her asymmetrical masks still attract unwanted attention and prevent her from fully participating in society, and the process of designing, building, and maintaining the masks is a huge challenge due to her chronic pain.

Design Process
1. Observation, Research, and Discussions About Ableism
2. Goal Setting
3. Identifying Design Requirements and Challenges
4. Prototype Iterations

Observation, Research, and Discussions
"Understanding the psychological state of the wearer can be immensely helpful for a wearable system."
Watkins and Dunne (131, 2016)

The inaccessible design of most physical environments is a common barrier for many PWD, yet the stigma associated with navigating that barrier isn't often addressed. So, the team began by discussing accessibility and disability stigma. Mia shared her experiences by telling them she was recently asked to leave a local grocery store because "she looked threatening in her dark leather mask," therefore, she requested that they make a new design that was clear and less conspicuous (Paget, pers. comm.). Inspired by science fiction, pop culture, and futuristic aesthetics, the team looked at a few different transparent materials (such as plastic) and store-bought face shields sometimes used in sports. "The plastic looked good. There have been more mask designs being made without dark colors and curves since COVID," said Mia (Paget, pers. comm.).

Goal Setting

While the masks Mia had created accurately expressed herself, they lacked the technology that could make them more comfortable and more reliable. The team's iterations addressed pain trigger mitigation (preventing exposure to sensory stimuli) by including protective barriers for sensitive nerve endings.

Identifying Design Requirements and Challenges

The team listed what their mask design would have to take into consideration: a snug fit, Mia's style, and therapeutic heat. They also wanted to ensure that the final product was comfortable enough to be worn for long periods of time, so they used the same rating scales used by most healthcare professionals to find and measure her pain points. The Numerical Rating Scale, Verbal Rating Scale, and Visual Rating Scale (Hjermstad et al. 2010) provided them with a subjective report of Mia's reactions, in addition to a series of interviews and material explorations.

Prototype Iterations

Material and Shape Exploration

Inspired by Mia's leather mask design, the first iteration featured a silhouette made with clear plastic and a variety of fabric adhesions, including cotton, wool, felt, and polyester. The team experimented with insulating foam. The final design work incorporated fleece because of its softness and ability to retain warmth.

Next, the team placed the fabric on top of the mask and traced the outer edges, eye openings, and mouth opening with a fabric marker, then cut it out. They laser cut computer-generated patterns and masquerade-inspired vectors onto the felt before carefully gluing it onto the clear mask with multipurpose adhesive and trimming the excess fabric.

Figure 4.114 Material swatches demonstrating the team's experiments with fabrics that have both softness and warmth retention. Courtesy of Open Style Lab.

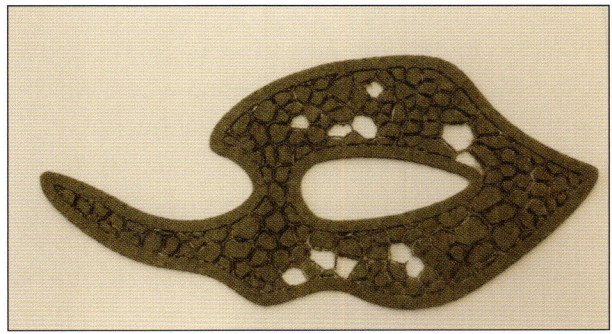

Figure 4.115 Mask iteration with laser cut details on felt. Courtesy of Open Style Lab.

Vacuum Forming Plastic

The team wanted to make a clear plastic mask tailored for Mia's face, so they turned to vacuum forming—a manufacturing process that uses heat at a pliable temperature to form a specific mold. Their process included taking 3D scans of her face, milling, and vacuum forming plastic. The team also consulted Jason Rizzo from Rogerson Orthotics & Prosthetics in Boston, who had experience creating custom-fit transparent facial orthotics.

In addition to their use for forming plastic, molds of Mia's face were used to test fabrics and design iterations when she wasn't present. After a few rounds, the team finalized the ideal form for their transparent mask with eye, mouth, and nostril openings.

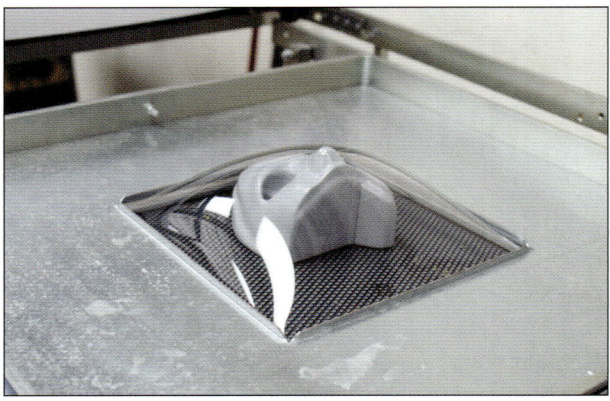 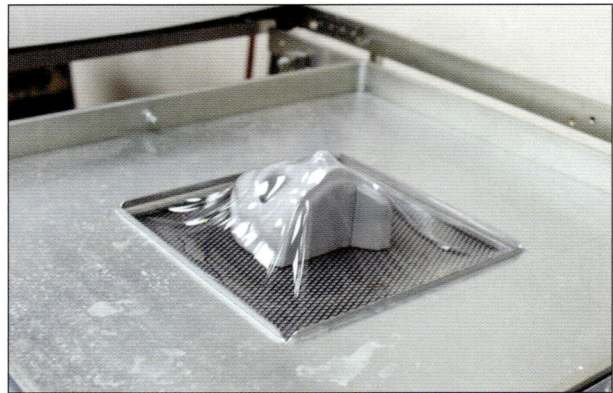

Figure 4.116a–b Vacuum forming process to create a custom-fit mask for Mia. Courtesy of Audrey Bosquet.

An initial vacuum-formed plastic mask provided a base for arranging resistive heating wires, which were then sealed after an additional layer of plastic was formed on top. Fabric liners were then applied to the inside layer using adhesive spray. Finally, design details were added, such as elastic straps for the head with a clasp closure sewn in to make the mask adjustable.

Wire Connection
Cuprothal® wires connected to a battery pack were used for optimal heat distribution and aesthetics. The heat is generated from resistive heating wire embedded between two layers of plastic vacuum formed into a mold of Mia's face. The goal was to use a battery pack that provides heat to be controlled by a smartphone using Bluetooth® and thermometer sensor feedback. The team used a thin wire called Cuprothal® because it would be less visible and was easier to embed between layers of plastic in a vacuum forming process.

Mia also investigated a special type of solder and wire wraps that could strengthen the connection at the mask interface; however, fixing the connection proved to be time-consuming and unreliable. Moreover, the heating wire was very thin and would break where it exited the mask because it was no longer protected.

Finally, the team noticed Mia sometimes had difficulty adjusting her plastic mask to fall in place

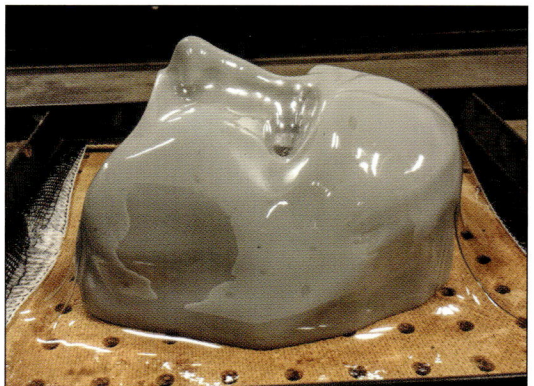 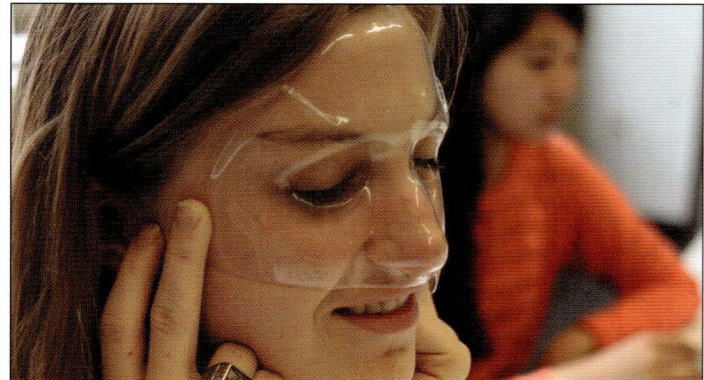

Figure 4.117a–b Final vacuum forming experiment with gray mold, worn by Audrey Bosquet. Courtesy of Open Style Lab.

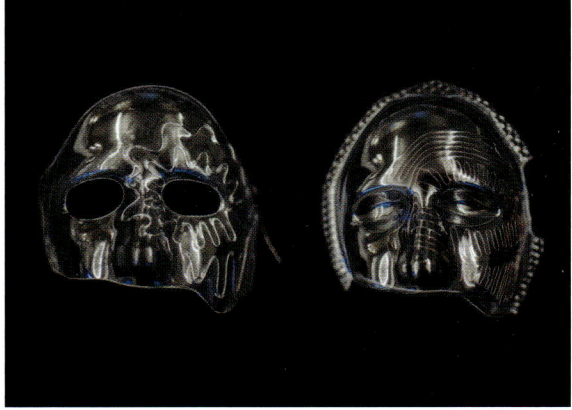

Figure 4.118a–b Microcontroller attached to a battery pack to provide heat to mask design. Two mask iterations created to compare wire thickness and style. Courtesy of Audrey Bosquet.

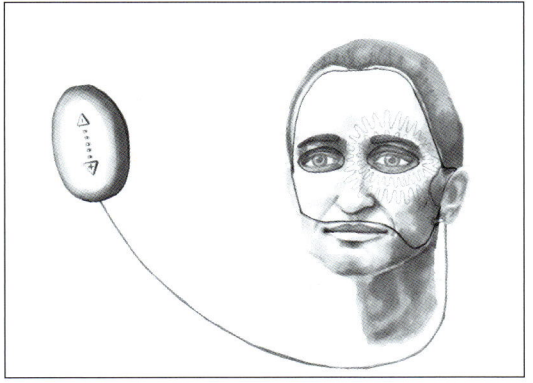

Figure 4.119a–b Illustration of a microcontroller connected to an external controller by Audrey Bosquet (courtesy of Audrey Bosquet). Final mask iteration exploring ideal wire thickness (Courtesy of Open Style Lab.).

and fit properly. Because adjusting the plastic mask was tiresome, the team created two types of masks: one that incorporated heat with wires and one that featured fabric to retain warmth.

Materials
- Laser-cutting machine
- Thermoforming machines
- Plastic sheets for vacuum molding
- Fabric (felt, cotton, interfacing, wool, leather)
- Plaster and resin for face molds
- 3M™ Super 77™ multipurpose spray adhesive
- Fabric elastic straps
- Clasp closures
- Arduino
- Battery pack
- Cuprothal® wires

Final Design Outcome

The team's designer, Audrey, commented that the biggest part of the process was iterating due to the many obstacles and technical constraints along the way. For example, the team scanned Mia's face to obtain a digital mesh for fabricating vacuum-forming molds. However, the scanning technology they had access to required multiple prototypes to determine the appropriate size. The final mold of her face was created 10 percent narrower to counteract the plastic's malleability and help it better fit Mia's face.

Case Studies, Stories, and Interviews

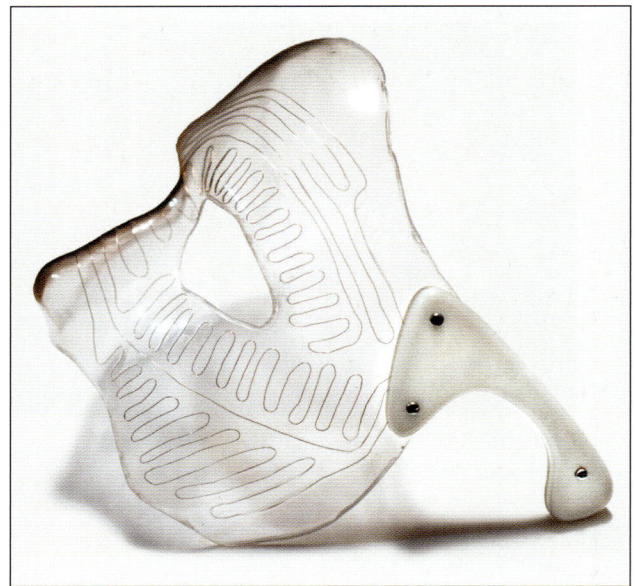

Figure 4.120a–b Final design of the Avisly mask. Courtesy of Audrey Bosquet.

"The best thing about this process was having other people work with me and trying to help me. I was being asked questions and included in everything," said Mia. "While some objectives of this project could not be completed or refined within the time constraints of the project, I was able to adapt and use the plastic molded masks for quite some time, increasing my functionality and improving my quality of life," explained Mia (Paget, pers. comm.). She also noted that creating something for a specific need that could later be applied to more people is the very definition of an inclusive process. Creating anything with a customized process often requires a specific and sometimes expensive tool—in this case, a vacuum-forming machine. Fortunately, today **DIY (do-it-yourself)** desktop vacuum-forming options are becoming more available.

References

Hjermstad, Marianne Jensen, Peter M. Fayers, Dagny F. Haugen, Augusto Caraceni, Geoffrey W. Hanks, Jon H. Loge, Robin Fainsinger, Nina Aass, Stein Kaasa, and European Palliative Care Research Collaborative (EPCRC). 2010. "Studies Comparing Numerical Rating Scales, Verbal Rating Scales, and Visual Analogue Scales for Assessment of Pain Intensity in Adults: A Systematic Literature Review." *J Pain Symptom Management* 41 (6): 1073–1093. doi:10.1016/j.jpainsymman.2010.08.016. PMID: 21621130 (accessed August 14, 2023).

LIULID
Ada Stewart, Julie Osipow, Heeyoung Kim, Grace Wu

Figure 4.121 Team LIULID from left to right: Grace Wu, Julie Osipow, Ada Stewart, and Heeyoung Kim. Courtesy of Open Style Lab.

Key Functional Features
Connectivity
Protection
Thermoregulation
Comfort

Summary
LIULID (Light It Up, Let It Drop) are athleisure wear pants that incorporate interactive technology to ease the act of dressing and encourage physical activity.

Design Goal
To create an everyday accessible garment that combines style and comfort while accommodating mobility, body temperature, and cleanliness.

Team LIULID
Ada Stewart: Client and subject-matter expert
Julie Osipow: Physical therapist
Heeyoung Kim: Design researcher
Grace Wu: Textile designer

Collaborator Acknowledgments
Parsons School of Design, Open Style Lab, The Riverside Premier Rehabilitation & Healing Center

Background
Presentation and self-expression are both parts of the daily fashion experience. However, the everyday portrayal of older bodies, particularly older women, is severely lacking. "Social invisibility arises from the acquisition of visible signs of aging," which assumes that older people don't desire stylish, colorful, or bright clothes (Clarke 2008). In reality, they do. And by combining fashion and function, adaptive design has the power to help older individuals live longer, happier, and more independent lives. For this collaboration, the senior client, Ada, helped guide the interdisciplinary team's strategic exploration and experimentation. What resulted was the creation of LIULID, an innovative pair of pants that was designed to (1) decrease the time and effort needed to dress and undress, (2) prioritize a desired hygienic style, and (3) encourage movement to optimize independence and the wearer's quality of life.

Introduction

Ada Stewart is a senior who lives in New York City, where she currently resides in The Riverside Premier Rehabilitation & Healing Center—a facility for short-term rehabilitation and long-term nursing care. Ada experiences several health factors that influence her daily routine, such as COPD (chronic obstructive pulmonary disease), RA (rheumatoid arthritis), bilateral DVT (deep vein thrombosis), and diabetes. For example, Ada experiences fatigue when exerting her body because she often requires supplemental oxygen to function. Inflammation from RA has significantly impacted the use of her dominant index finger, dominant hand, and shoulder joints, and, possibly because of her DVT, she has edema in both legs, which causes swelling from the excess fluid trapped in the body's tissues.

Ada has regular access to healthcare staff and uses a wheelchair for transportation. When asked why she wanted to participate in Open Style Lab's program, she replied, "To show what we [older adults] need. Clothes are important to us" (Stewart, pers. comm.). During the team's first interview with Ada, they learned that while she experiences pain in her upper extremities when tying her nightgown, she doesn't let it interfere with her love of crocheting.

Design Process
1. Research
2. Observation and Interview
3. Goal Setting
4. Journey Map
5. Identifying Design Requirements and Challenges
6. Prototype Planning and Development

Design Requirements

The team wanted to create a garment that (1) lets Ada feel clean, comfortable, and presentable and (2) helps her maintain her independence when dressing (The American Occupational Therapy Association 2020). Therefore, they explored a pant design with her that would:

1. support the efficiency of dressing by reducing the time it takes to don and doff the pants;
2. utilize textiles and materials that support the regulation of body temperature; and
3. provide garment cleanliness and the preservation of the wearer's dignity and confidence.

Research

Older adults with multiple diagnoses often struggle with carrying out ADLs, such as eating, dressing, and practicing personal hygiene. Although there are products available for purchase that help one maintain their independence, they're usually medical devices that are dismissed because of perceived opinions and assumed stigmas around using assistive equipment (Resnik 2009). Without technology designed to help them accomplish ADLs, this generation typically has no choice but to employ a personal caregiver or move to an assisted living facility.

Observation and Interview

Members of the team observed Ada's daily activities. Gradually, the team learned how Ada gets dressed, her opinions on existing clothing, and some of her hobbies, such as crocheting.

Objective measures were assessed using task analysis quantified by time, standard muscle testing for strength, and range of motion. Then, subjective data was collected to help the team better understand Ada's status of independence, perceived capabilities, personality, and personal values. Together, these methods helped identify a daily task that posed a considerable challenge both quantitatively and qualitatively—putting on pants. This eight-minute process required a lot of

Figure 4.122 Julie Osipow, Ada Stewart, and Heeyoung Kim discussing various clothing styles and materials. Courtesy of Open Style Lab.

energy and often caused fatigue and breaks. It consisted of:

- rolling up each leg to shorten the fabric, making it more manageable, preventing it from touching the floor, and keeping it clean before wear;
- pulling the pants over each leg individually, which requires significant trunk flexion to both hold the garment and insert the correct foot inside the correct leg; and
- standing up to pull the pants over her thighs and hips before returning to a seated position.

Learning about these specific challenges inspired the team to explore origami, the art of folding paper, and ultimately led to the creation of the LIULID pants (see fig. 4.123a–b).

Prototype Iteration 1
Origami Exploration
Inspired by the art of origami, the team decided to incorporate pre-established pleats, or folds, that could help decrease the time it takes a wheelchair user to don a pair of pants with less physical exertion. Using wool (the chosen fabric) that could withstand multiple washes, they realized they had to make the folds permanent. So, Team LIULID experimented with two different origami tessellations (patterns that repeat

Figure 4.123a–b Material exploration of pleats and fabric swatches for pant design. Brown origami paper exploration created by Grace Wu.
Courtesy of Open Style Lab.

Figure 4.124 Origami paper exploration in different materials created by Grace Wu and Heeyoung Kim.
Courtesy of Open Style Lab.

themselves). Their first attempt was implementing the Original Miura-Ori fold by Koryo Miura (see fig. 4.123a–b) by using a laser cutter to engrave creases into kraft paper. Then they

Case Studies, Stories, and Interviews

molded the shape of the fabric by sandwiching it between two paper models that folded. Securing each fold with a tying mechanism and ironing the paper model, they were able to make pleats. Unfortunately, these pleats were not retained after being washed, and the rigidity of the origami fold hindered the ability of the fabric to form into the correct shape for a pant leg.

Prototype Iteration 2
3D Printing
After exploring 3D printing options, the team discovered it was challenging to manipulate various materials to hold an intended shape because fabric isn't structurally like paper—it lacks rigidity. Since using a more structured fabric could potentially compromise the garment's level of comfort, they chose to maintain their original fabric choice—a wool from Woolmark that had TPU (thermoplastic polyurethane) bonded on one side—and experimented with different ways to secure the pleat technique. Then, the TPU side inspired them to try 3D printing on fabric using a TPU filament, allowing for easy adhesion and laundering.

Next, they created accordion fold designs on the pants. They printed horizontal panels using NinjaFlex® TPU on one side of the wool to maintain

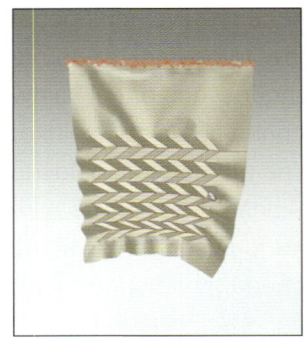 

Figure 4.126 Digital rendering using Clo3D software that helped depict the pattern movement on fabric. Images created by Grace Wu and Heeyoung Kim. Courtesy of Open Style Lab.

the structure of its pleats. By adding slits to each panel, they created smaller blocks to give the material more movement, which let the 3D printed fabric curve to the shape of a leg (see fig. 4.125).

Prototype Iteration 3
Pulley System
After establishing a technical way to fold the fabric into a manageable form, the team ideated a way for the user to efficiently collapse and expand the folds with minimal effort. Keeping in mind Ada's dexterity, strength, and areas of pain, they then developed a pulley system that consisted of an elastic cord that attached to the 3D printed portion of the pant leg (see fig. 4.128a–c) and the waistline. This lets Ada use her hands, fingers, or wrists to pull the elastic loop, giving her more leverage to collapse the length of

Figure 4.125 3D printing on fabric exploration created by Grace Wu and Heeyoung Kim. Courtesy of Open Style Lab.

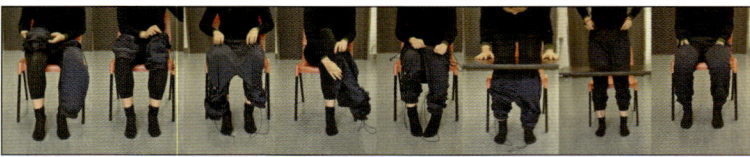

Figure 4.127 Dressing process of LIULID demonstrated by Heeyoung Kim. Courtesy of Heeyoung Kim.

Figure 4.128a–c Dressing pulley system prototypes created by Heeyoung Kim. Courtesy of Open Style Lab.

the LIULID garment and complete the process of dressing while standing. Having identified the latter activity as the most physically challenging one for Ada—sometimes requiring two or three tries, the team wanted to decrease the time and effort she spent standing to pull up her pants and increase the use of her upper extremities.

Prototype Iteration 4
Cohesive Look
Athleisure, which combines the comfort and aesthetics of athletic wear with everyday fashion, has become a popular trend over the last two decades (Salonga 2018). Sneakers, leggings, sweatshirts, and sweatpants have infiltrated the collections of high-end designers like Gucci and practically defined the success of activewear brands like Athleta and Lululemon. Ada's closet mainly consisted of casual attire made from stretch or cotton fabrics, which informed the team's decision to create a pair of pants influenced by athleisure style.

Prototype Iteration 5
Integrating POF Textiles
Physical activity is necessary to maintain muscle tone, decrease pain, increase endurance, regulate body temperature, improve circulation, and benefit one's metabolism (Anton et al. 2015). Because exercise was important to Ada, the team incorporated a textile strip with polymeric optical fiber (POF) to create sensors within the pant design. Without compromising the integrity of the textile technology, they were easy to attach and remove from the pants for proper laundering, and on days when being active wasn't an option due to Ada's health conditions. The sensors within the textiles were aimed to light up when Ada's knees would bend in a particular degree, signifying she has correctly performed her physical therapy exercise. While the team did not have time to integrate the textile fully into the final garment design, the goal was to help Ada avoid muscle atrophy when using her wheelchair.

Project Materials
- Woolmark
- Polartec®
- Various wool blends
- Cotton
- Paper and scissors (for origami tests)
- 3D printing filament
- 3D printer

Final Design Outcome
Team Ada created the LIULID pants, which incorporated pulleys into the waistline to help Ada pull them up while standing by using her upper extremities. These successfully reduced the time and effort it took her to get dressed. The garment also featured the textile strips to encourage movement and lap pockets lined with Polartec® to keep Ada's hands warm, help regulate her body temperature, and reduce her tendency to compromise her posture for the sake of warmth.

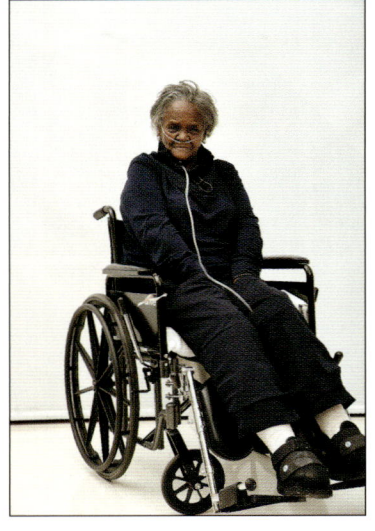
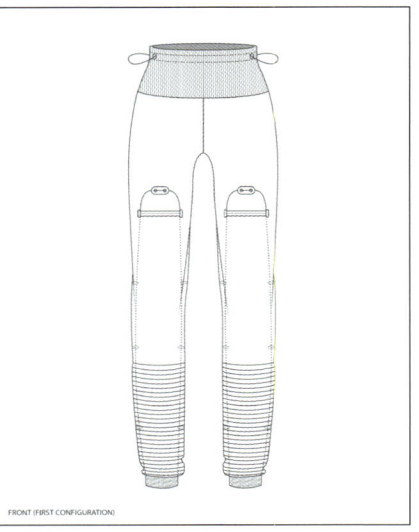

Figure 4.129 Ada Stewart wearing navy-colored LIULID pants. Courtesy of Open Style Lab.

Figure 4.130 Illustration of LIULID pants for standing and seated body postures drawn by Heeyoung Kim. Courtesy of Open Style Lab.

An elastic cord was threaded through the fabric using metal eyelets on each TPU plastic panel to reduce friction and decrease the force needed to collapse the pant leg. When the garment was put on, the cinched hem tapered to the leg automatically expanded the pants without any extra effort from Ada to unfold the pleat.

References

Anton, Stephen D., Adam J. Woods, Tetso Ashizawa, Diana Barb, Thomas W. Buford, Christy S. Carter, David J. Clark, Ronald A. Cohen, Duane B. Corbett, Yenisel Cruz-Almeida, Vonetta Dotson, Natalie Ebner, Philip A. Efron, Roger B. Fillingim, Thomas C. Foster, David M. Gundermann, Anna-Maria Joseph, Christy Karabetian, Christiaan Leeuwenburgh, Todd M. Manini, Michael Marsiske, Robert T. Mankowski, Heather L. Mutchie, Michael G. Perri, Sanjay Ranka, Parisa Rashidi, Bhanuprasad Sandesara, Philip J. Scarpace, Kimberly T. Sibille, Laurence M. Solberg, Shinichi Someya, Connie Uphold, Stephanie Wohlgemuth, Samuel Shangwu Wu, and Marco Pahor. 2015. "Successful Aging: Advancing the Science of Physical Independence in Older Adults." *Ageing Research Reviews* 24: 304–327. doi:10.1016/j.arr.2015.09.005 (accessed August 14, 2023).

Clarke, Laura Hurd and Meridith Griffin. 2008. "Visible and Invisible Ageing: Beauty Work as a Response to Ageism." *Ageing and Society* 28 (5): 653–674. doi: 10.1017/s0144686x07007003 (accessed August 14, 2023).

Resnik, Linda, Susan Allen, Deborah Isenstadt, Melanie Wasserman, and Lisa Iezzoni. 2009. "Perspectives on Use of Mobility Aids in a Diverse Population of Seniors: Implications for Intervention." *Disability and Health Journal* 2 (2): 77–85. doi:10.1016/j.dhjo.2008.12.002 (accessed August 14, 2023).

Salonga, Bianca. 2018. "Chic Athleisure Labels To Motivate An Active Lifestyle." *Forbes*, July 15. www.forbes.com/sites/biancasalonga/2018/07/15/chic-athleisure-labels-to-motivate-an-active-lifestyle/#4b1baae218bd (accessed August 11, 2018).

The American Occupational Therapy Association. 2020. "Occupational Therapy Practice Framework: Domain and Process Fourth Edition." *American Journal of Occupational Therapy* 74 (2): 1–87. doi:10.5014/ajot.2020.74S2001 (accessed August 14, 2023).

Warmed Bomber
Silvo Mehle, Jensin Okunishi, Evrim Buyukaslan Oosterom, Petja Zorec

Key Functional Features
Connectivity
Protection
Thermoregulation

Summary
The Warmed Bomber is an electrically heated jacket with temperature control using Bluetooth®. It features elastic armholes for easier dressing, a retractable hood, an extended back hem for full coverage in the seated position, and sleeve cuffs that double as mittens.

Design Goal
To create a custom-fit jacket that supports thermoregulation and easier dressing features for people experiencing spasticity and limited range of motion.

Team Warmed Bomber
Silvo Mehle: Client, artist, and subject-matter expert
Jensin Okunishi: Textile designer
Evrim Buyukaslan Oosterom: Textile engineer
Petja Zorec: Fashion designer

Collaborator Acknowledgments
RogLab and Open Style Lab co-hosted the DESIGN (DIS)ABILITY workshop in Ljubljana, Slovenia, from March 30 to April 3, 2015, which included Meta Stular, Dr. Grace Teo, Tea Pristolic, William Li, Sanja Grcic, and Tomo Per.

RogLab is a Ljubljana-based production space and public facility dedicated to experimentation and the exchange of ideas in the field of design, architecture, visual arts, and new technologies. RogLab's programs focus on offering production tools and creative uses of 3D technologies, enabling interdisciplinary collaboration as well as research between creative activities and business. Many of their projects address pressing issues in urban environments and current challenges in architecture and design with an emphasis on social and environmental responsibility.

Background
Like architecture, fashion is represented in different scales and forms of shelter and explores the relationships between function, aesthetics, and comfort. Clothing acts "as a package for the body or textile architecture" (Giesel 2012). It is the interface that negotiates between the self and the environment. Clothing, therefore, becomes a tool for communication and can create new meanings for those who view, model, and wear it. Outerwear can contribute to the maintenance of the body's thermal balance while making a fashion statement, as seen in athleisure. This is because its fabric usually provides better breathability and patternmaking techniques that privilege ventilation in targeted areas of the body, such as sweat glands. For example, Bomber jackets, also known as flight jackets during the First World War, were designed to keep pilots warm in uninsulated planes.

This design process explores a bomber jacket as an accessible and functional garment for disability needs. As part of the ongoing collaboration for DESIGN (DIS)ABILITY between RogLab and Open Style Lab, the bomber jacket project was held in Ljubljana with the goal of creating fashion accessories with and for users with physical disabilities. The team was challenged to create both aesthetically pleasing and functional fashion designs while receiving mentorship from an interdisciplinary team of experts from both design and engineering backgrounds. This let the participants exercise their skill sets, master a variety of fabrication techniques, and learn basic programming and electronics for wearable technology.

Introduction
Silvo is an accomplished mouth artist from Slovenia who creates still-life and landscape paintings. He has tetraplegia, a paralysis that affects both arms and legs, that was caused by a traffic accident, so he uses an electric wheelchair that he operates with a joystick positioned near his face. His partner Helena often helps him get dressed—particularly during the colder months when he needs more clothing. Staying warm is essential to his well-being; therefore, Silvo's concerns included feeling cold around his kidneys, his inability to find gloves that fit his bent fingers, and his dislike for his few jacket options—all of which featured synthetic materials that irritate his skin. Lastly—and most importantly—he wanted a jacket that fit his body properly when he was seated.

The Project
- Technology: sewing, knitting, soldering
- Materials: waxed cotton, wool trims, metal zipper, wearable electronics, Bluetooth® device
- Size of garment prototypes: Large (L)

Design Process
1. Observation and Research
2. Goal Setting
3. Identifying Design Requirements
4. Prototype Iterations

Observation and Research
"It isn't easy to estimate what a disabled person may need in terms of the stuff we use daily, including clothing, since every disability is unique, and each person has special needs."
  Evrim Buyukaslan Oosterom (pers. comm.)

The team began by investigating personal biases and Silvo's experiences with his disability before constructing any designs. They watched how Silvo got dressed, which often included Helena helping him don long-sleeved shirts and jackets by placing his hands through the armholes, then flipping the garment over his head to pull the rest of the material behind him to cover his back. This process was not easy, causing Silvo to strain his neck and Helena to exert a lot of physical strength.

"Speaking with Silvo gave us a window into his personal style and influenced the aesthetic of our design. And observing how he interacts with his caretaker helped us better understand the things that he can do on his own versus what he needs help with."
  Jensin Okunishi (pers. comm.)

Afterward, the team researched existing jacket solutions that provide heat. They discovered that motorcycle jackets with fabric patches can run an electric current through a high-resistance electrical wire and heat up its wearer. However, they also found cases of heated jackets worn by wheelchair users that caused burns—the reason being that being placed up against the back upholstery didn't allow for airflow, causing the garment to overheat. So, the team investigated safety features for their design before synthesizing their learnings.

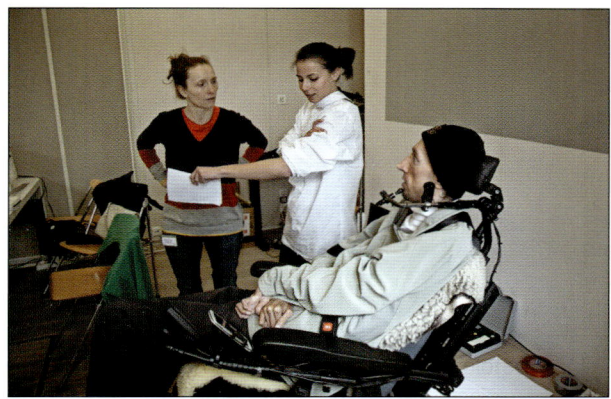

Figure 4.131 Petja Zorec, Helena, and Silvo Mehle discussing designs that could best cover Silvo's arm.
Courtesy of Rog Centre, photo: Manca Juvan.

Figure 4.132 Computer sketch of possible jacket pattern design created by Jensin Okunishi.
Courtesy of Open Style Lab.

Figure 4.133 Sketches of bomber jacket style.
Drawing by Petja Zorec.
Courtesy of Rog Centre, photo: Manca Juvan.

Goal Setting

When the team asked to see the types of jackets Silvo already wore, Helena shared with them three different silver ones—all of which he wasn't fully satisfied with. They could immediately tell that they didn't cover his hands, were difficult to put on, and lacked rain protection. From there, together with Silvo, they scored five different metrics of success: comfort, warmth, style, fit, and ease of dressing.

Design Requirements
- Comfortable: soft and non-irritating
- Warm: quickly provides heat all around the torso
- Stylish: aesthetically pleasing with athletic or sporty design elements
- Fitted: shorter in the front and longer in the back to cover a seated body
- Easy to don: reduces the time it takes the wearer and caretaker to put on the jacket

Prototype 1

The team wanted to make sure that the garment they created would be aesthetically pleasing, so they asked Silvo to share pictures of the types of jackets he liked. Many of them featured a quilted design and a thin, sleek collar in different shades of gray. Based on these references, Jensen digitally designed a quilted pattern with different accents, including green thread.

Then, Petja designed patterns based on Silvo's existing jackets and created pieces of the garment in muslin for fit and style. Aesthetic considerations that aligned with Silvo's identity were explored in the following design features: ribbed knit armholes, a sleek retractable hood that can be removed from the collar and elongated, and ribbed knit sleeve cuffs.

The definition of comfortable includes a person's convenience and well-being. Looking at apparel as a habitat for the body, the team experimented

Case Studies, Stories, and Interviews

Figure 4.134 Petja Zorec cutting out fabric based on pattern design.
Courtesy of Rog Centre, photo: Manca Juvan.

with various silhouettes and fits for sleeves to better gauge Silvo's level of comfort—in turn, applying ergonomics into fashion design. When clothes fit the user, the result can be more comfortable, produce higher productivity, and reduce stress (ELsayed et al. 2019).

Prototype 2
The materials used were essential to the success of this project; however, sourcing them proved to be a main challenge at first. The team experimented with different fabrics that could be carefully placed with heating electronics inside the jacket, including polyurethane leather, water-resistant technical fabrics, and soft, stretchy boucle materials.

They also found a waxed cotton with finishes that offered some protection from cold and rainy weather and, finally, a heat-conductive silver microfiber to cover the pads in the lining. A stretchy ribbing in dark gray and black was then used to accent the sleeves and arm areas to provide extra give when putting on the jacket. While the team wanted to initially create mittens or gloves for Silvo, they were surprised to find that the ribbed cuffs of the sleeves already covered his hands sufficiently.

Prototype 3
"Designing a smart garment that's inclusive of disabilities requires a lot of different expertise. Without any one of us, the project could have potentially had a really different outcome."
                                    Jensin Okunishi (pers. comm.)

One of the most challenging elements of this design process was integrating technology that already complemented Silvo's devices and wheelchair. For the jacket's third prototype, the team found a thrifted motorcycle jacket that had heating pads and a battery pack attachment. They made sure to test the heat levels of these pads before integrating them into the garment design by placing them against themselves. While they provided instant warmth, the battery pack proved

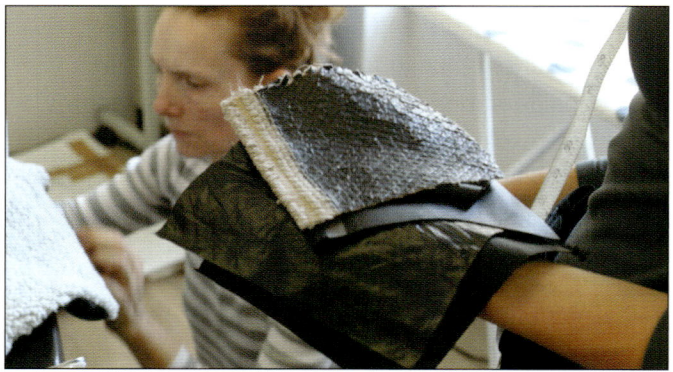 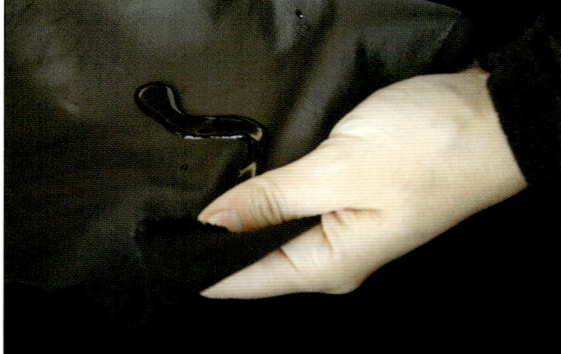

Figure 4.135a–b Black fabric swatches with water-resistive properties. Courtesy of Open Style Lab.

Figure 4.136 Evrim Buyukaslan Oosterom experimenting with heated fabric pads. Courtesy of Open Style Lab.

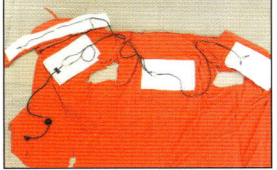

Figure 4.137a–b Microcontroller and battery pack circuit design exploration created by Jensin Okunishi with guidance from Grace Jun and William Li. Exploration of wire and heating pad placements on a red-colored fabric. Courtesy of Open Style Lab.

Figure 4.138a–b Final design of Warmed Bomber details. Courtesy of Open Style Lab.

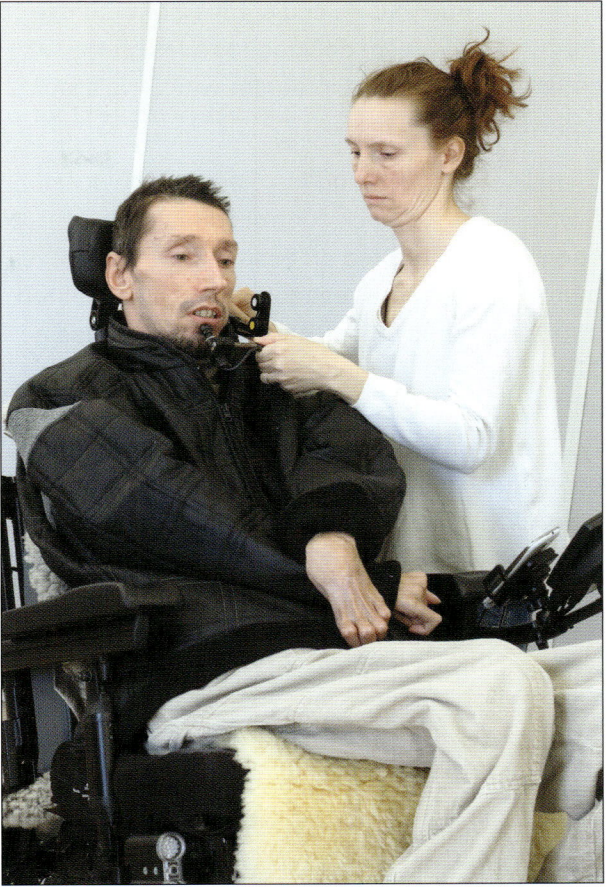

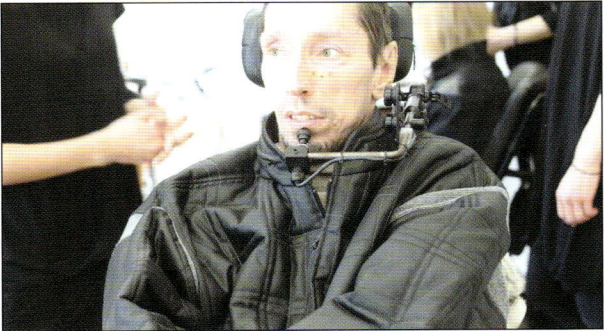

Figure 4.139a–b Silvo Mehle wearing Warmed Bomber final design. Courtesy of Open Style Lab.

Case Studies, Stories, and Interviews

to be too large, heavy, and cumbersome to use. Therefore, the team decided to redesign the battery pack to better integrate with Silvo's jacket and wheelchair.

In addition to operating his electric wheelchair, Silvo's mouth joystick controlled his mobile phone, which offered an essential way to manage even more technology. He also pointed out that electric wheelchairs have two times a 12-volt accumulator battery and don't need an additional battery for an accessory, just a plug. So, the team created a Bluetooth® device that connected to a detachable wire that fed into his jacket—this way, he could use his phone to turn on and off the heating. The first iterations were connected to an Arduino, a microcontroller board, and experimented with a larger **circuit** design, as referenced in figures 4.137a–b.

Final Design
This sporty, stylish men's bomber jacket was designed to be accessible for people living with tetraplegia. It's made of waxed cotton for protection against the cold and rain and features functional elements that include a cropped front, looser silhouette for the arms and shoulders, ribbed collar, waistband, and cuffed sleeves.

The wearable electronics integrated between the lining and the filler keep the back area warm, with the electric wheelchair battery serving as the energy source. And donning and doffing are ultimately made easier with a stretchable trim in the shoulder area.

References

Anton, Stephen D., Adam J. Woods, Tetso Ashizawa, Diana Barb, Thomas W. Buford, Christy S. Carter, David J. Clark, Ronald A. Cohen, Duane B. Corbett, Yenisel Cruz-Almeida, Vonetta Dotson, Natalie Ebner, Philip A. Efron, Roger B. Fillingim, Thomas C. Foster, David M. Gundermann, Anna-Maria Joseph, Christy Karabetian, Christiaan Leeuwenburgh, Todd M. Manini, Michael Marsiske, Robert T. Mankowski, Heather L. Mutchie, Michael G. Perri, Sanjay Ranka, Parisa Rashidi, Bhanuprasad Sandesara, Philip J. Scarpace, Kimberly T. Sibille, Laurence M. Solberg, Shinichi Someya, Connie Uphold, Stephanie Wohlgemuth, Samuel Shangwu Wu, and Marco Pahor. 2015. "Successful Aging: Advancing the Science of Physical Independence in Older Adults." *Ageing Research Reviews* 24: 304–327. doi:10.1016/j.arr.2015.09.005 (accessed August 14, 2023).

ASIA (American Spinal Injury Association). 2022. *Adaptive Clothing for People with Spinal Cord Injury: Where Function Meets Fashion.* Abstract 14. New Orleans, LA.

Bezyak, Jill L., Scott A. Sabella, and Robert H. Gattis. 2017. "Public Transportation: An Investigation of Barriers for People With Disabilities." *Journal of Disability Policy Studies* 28 (1): 52–60. doi:10.1177/1044207317702070 (accessed August 14, 2023).

Christopher & Dana Reeve Foundation. 2023. "What is a Complete vs Incomplete Spinal Cord Injury?". https://www.christopherreeve.org/todays-care/living-with-paralysis/newly-paralyzed/how-is-an-sci-defined-and-what-is-a-complete-vs-incomplete-injury/, February 20 (accessed October 26, 2023).

Clarke, Laura Hurd and Meridith Griffin. 2008. "Visible and Invisible Ageing: Beauty Work as a Response to Ageism." *Ageing and Society* 28 (5): 653–674. doi:10.1017/s0144686x07007003 (accessed August 14, 2023).

Core Jr. 2018. "Meet Julia Liao, the Winner of This Year's Core77 x A/D/O Residency." *Core77*, March 19. https://www.core77.com/posts/74996/Meet-Julia-Liao-the-Winner-of-This-Years-Core77-x-ADO-Residency (accessed August 14, 2023).

Dunn, Winnie. 1999. *Sensory Profile*. Amsterdam: Elsevier.

ELsayed, Wafaa A., Maha M.T. Eladwi, Nagah S. Ashour, Rania N. Shaker, Eman M.S. Shaheen. 2019. "Ergonomics Approach for Fashionable Apparel Design." *International Design Journal* 9 (3): 273–280. doi:10.21608/idj.2019.82831 (accessed August 14, 2023).

Giesel, Aline and Patricía de Mello Souza. 2012. "The Correlation between Thermal Comfort in Buildings and Fashion Products." *WORK: A Journal of Prevention Assessment & Rehabilitation* 41. doi:10.3233/wor-2012-0882-5561 (accessed August 14, 2023).

Hjermstad, Marianne Jensen, Peter M. Fayers, Dagny F. Haugen, Augusto Caraceni, Geoffrey W. Hanks, Jon H. Loge, Robin Fainsinger, Nina Aass, Stein Kaasa, and European Palliative Care Research Collaborative (EPCRC). 2010. "Studies Comparing Numerical Rating Scales, Verbal Rating Scales, and Visual Analogue Scales for Assessment of Pain Intensity in Adults: A Systematic Literature Review." *J Pain Symptom Management* 41 (6): 1073–1093. doi:10.1016/j.jpainsymman.2010.08.016 (accessed August 14, 2023).

Kaiser, Susan B., Carla M. Freeman, and Stacy B. Wingate. 1985. "Stigmata and Negotiated Outcomes: Management of Appearance by Persons with Physical Disabilities." *Deviant Behavior* 6 (2): 205–224, doi:10.1080/01639625.1985.9967670 (accessed August 14, 2023).

Kielhofner, Gary. 2008. *Model of Human Occupation: Theory and Application*, Fourth Edition. Philadelphia: Lippincott Williams & Wilkins.

Mahr, Dominik, Annouk Lievens, and Vera Blazevic. 2014. "The Value of Customer Co-Created Knowledge during the Innovation Process." *Journal of Product Innovation Management* 31 (3): 599–615. doi:10.1111/jpim.12116 (accessed August 14, 2023).

Mallon, Christina. 2018. "The Most Valuable Person You Haven't Hired Yet." *Fast Company*, July 9. https://www.fastcompany.com/90180550/the-most-valuable-person-you-havent-hired-yet (accessed August 14, 2023).

NSCISC (National Spinal Cord Injury Statistical Center). 2006. *2006 NSCISC Annual Report for the Model Spinal Cord Injury Care Systems*. Birmingham, AL: National Spinal Cord Injury Statistical Center. https://www.nscisc.uab.edu/PublicDocuments/reports/pdf/NSCIC%20Annual%2006.pdf (accessed July 29, 2017).

Pendleton, Heidi McHugh, and Winifred Schultz-Krohn. 2013. *Pedretti's Occupational Therapy: Practice Skills for Physical Dysfunction*, 7th ed. St Louis, MO: Elsevier.

Poon, Linda. 2015. "Why Are Wheelchair Users More Likely to Be Killed in Traffic Than Other Pedestrians?" *Bloomberg*, November 19, 2015. https://www.bloomberg.com/news/articles/2015-11-19/study-pedestrians-in-wheelchairs-are-a-third-more-likely-to-be-killed-in-road-accidents (accessed August 08, 2017).

Region 8 News. 2016. "Fashion w/ Special Needs." September 19, 2016. Video, 0:30 https://es-la.facebook.com/Region8News/videos/10153877266253148/

Resnik, Linda, Susan Allen, Deborah Isenstadt, Melanie Wasserman, and Lisa Iezzoni. 2009. "Perspectives on Use of Mobility Aids in a Diverse Population of Seniors: Implications for Intervention." *Disability and Health Journal* 2 (2): 77–85. doi:10.1016/j.dhjo.2008.12.002 (accessed August 14, 2023).

Salonga, Bianca. 2018. "Chic Athleisure Labels To Motivate An Active Lifestyle." *Forbes*, July 15. www.forbes.com/sites/biancasalonga/2018/07/15/chic-athleisure-labels-to-motivate-an-active-lifestyle/#4b1baae218bd (accessed August 11, 2018).

The American Occupational Therapy Association. 2020. "Occupational Therapy Practice Framework: Domain and Process Fourth Edition." *American Journal of Occupational Therapy* 74 (2): 1–87. doi:10.5014/ajot.2020.74S2001 (accessed August 14, 2023).

Tomchek, Scott D. and Winnie Dunn. 2007. "Sensory Processing in Children with and without Autism: A Comparative Study Using the Short Sensory Profile." *American Journal of Occupational Therapy* 61: 190–200. doi:10.5014/ajot.61.2.190 (accessed August 14, 2023).

Williamson, Bess. 2019. *Accessible America: A History of Disability and Design*. New York: NYU Press.

Wong, Sandy. 2018. "Traveling with Blindness: A Qualitative Space-Time Approach to Understanding Visual Impairment and Urban Mobility." *Health & Place* 49: 85–92. doi:10.1016/j.healthplace.2017.11.009 (accessed August 14, 2023).

Wong, Sandy, Sara L. McLafferty, Arrianna M. Planey, and Valerie A. Preston. 2020. "Disability, Wages, and Commuting in New York." *Journal of Transport Geography* 87. doi:10.1016/j.jtrangeo.2020.102818 (accessed August 14, 2023).

Chapter 5
Conclusion

Fashion is a vehicle for all people to express themselves. With a $2.5 trillion global revenue each year, it is one of the world's largest economic industries (Cabigiosu 2020). Yet the entry of adaptive clothing into the fashion mainstream has not been without its issues. From inaccessible stores to marketing campaigns that neglect disability visibility, there are still many stages of the fashion process that can be more inclusive. With such global influence, fashion brands have a financial opportunity, and social responsibility, to shift the narrative by making PWD more visible in the industry. A lack of disability visibility is a contributing factor limiting adaptive fashion from being integrated into mainstream culture. The other factor is scaling adaptive design products and business models.

Increasing Visibility

Podcast stories like Fashionably Tardy, co-founded by Natalie Trevonne, and disabled models like Jillian Mercado are some of the ways adaptive fashion is gaining greater visibility. Adaptive fashion encourages collaboration between disabled and nondisabled people, which increases creativity in product branding and marketing. For example, marketing adaptive fashion helps identify design exclusion by avoiding ableist language and questions the idealized body or normalization of dominant body types. Partnerships and collaborations are one of the ways PWD can be represented across all stages of the fashion industry. Christin Siriano's collaboration with Selma Blair on an adaptive line in 2018 was prompted by Blair's challenges to find clothes that accommodate her needs after being diagnosed with MS (Farra 2021). Internationally, Sinead Burke partners with Gucci "to co-lead a project to create employment opportunities for disabled people at the company" (Rawlins 2021).

More and more brands are embracing diversity and accessibility. Recognizable brands such as Kohl's and Nike are venturing into the adaptive space as part of their business strategy. Chromat's body-inclusive swimwear in the Chromat x Tourmaline collection is inclusive of "girls who don't tuck, trans femmes, non-binary and transmasculine people who pack, intersex people, women, and men" (Chromat 2021). Chromat introduces a range of swimsuits designed for trans women with convertible pieces that can be worn by PWD in the 2022 show, as depicted in figures 5.1–5.2.

Fashion shows are another way disability representation can be achieved. Organizations like Runway of Dreams showcase disabled models

Figure 5.1 Jasmine Guzman, Jordyn Harper, and Helen Peña in Chromat x Tourmaline SS22. Photography by Hatnim Lee. Image courtesy of Hatnim Lee.

Figure 5.2 Josh Allen, Ericka Hart and Maya Finoh in Chromat x Tourmaline SS22. Photography by Hatnim Lee. Image courtesy of Hatnim Lee.

Figure 5.3 Image of Kiley Gallant, Sawsan Zakaria, James Ian, Jenna Dewar, Joe Lakhman, Hannah Burcaw, and Shane Burcaw photographed by Shutterstock. Image provided by Genentech.
Image courtesy of Andrew H. Walker, Shutterstock

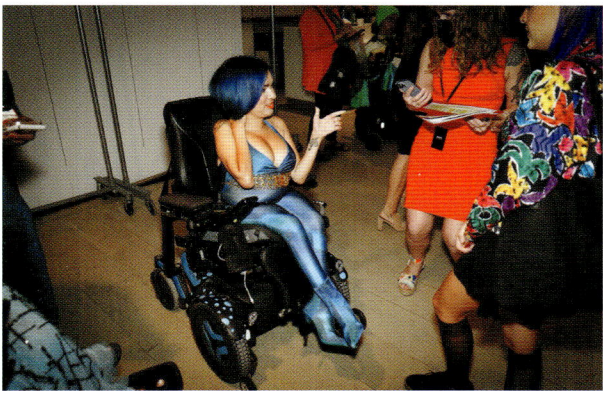

Figure 5.4 Image of Sawsan Zakaria photographed by Clifton Mooney Photography.
Image courtesy of Genentech.

during New York Fashion Week (Kaufman 2022). Additionally, "biotechnology company Genentech and Open Style Lab came together with the SMA (spinal muscular atrophy) community to create Double Take, a first-of-its-kind fashion show featuring members of the community modeling adaptive clothing designs created for them by OSL (Open Style Lab) fellows" (Criales-Unzueta 2022).

From nonprofits to pharmaceutical companies, collaborations are essential to moving design practices into a more inclusive space. Companies furthering scientific innovations and pharmaceutical advancements like Genentech demonstrate such collaborations are not only possible but essential. For example, Genentech launched their SMA My Way program in 2020, which is a collaboration with the SMA community that aims to support the community as a whole. Senior Director of Marketing at Genentech, Michael Dunn, explains how adaptive fashion is part of a holistic lifestyle, which echoes Genentech's corporate commitment toward DE&I (diversity, equity, and inclusion) by listening and supporting the SMA community and broader disability communities:

"Making treatments is just one of the responsibilities we have. We always look to listen and partner with the communities that we serve, to understand their needs beyond treatment. The common theme that we kept hearing from members of the SMA community is that fashion is a way of self-expression and self-identity. Therefore, a lack of accessible adaptive fashion prohibits disabled individuals to be able to express themselves."

Dunn (2022)

This observation led to a collaboration between Genentech and Open Style Lab to raise disability visibility during New York Fashion Week in the fall of 2022. Genentech's example boldly champions important aspects of the SMA community through external programming and corporate commitment, which has set an example for other companies to follow, such as tech or CPG (consumer packing goods) companies.

Broadening the adaptive fashion marketplace by highlighting accessible products made by disabled designers are US companies like Rebirth Garments and Social Surge. A community of nondisabled and disabled people, Social Surge is an apparel brand focused on universal and human-centered design and was co-founded by a person with a disability, Meredith Wells. Social Surge is centered on accessibility for everybody—specifically people

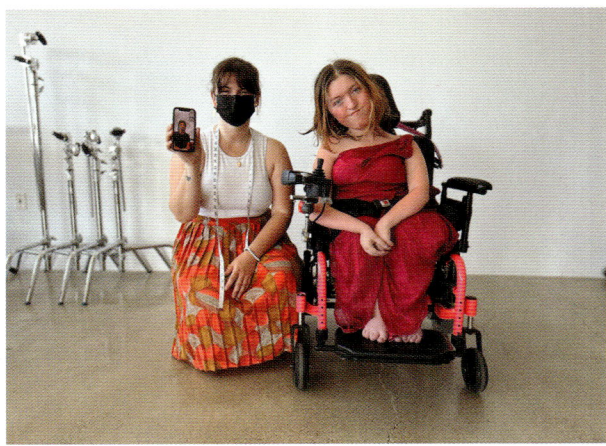

Figure 5.5 Image of Kellie Cusack and Kiley Gallant photographed by Syneos Health.
Image courtesy of Genentech.

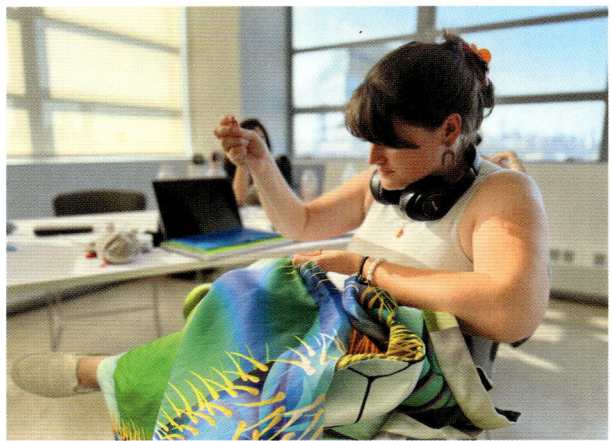

Figure 5.6 Image of Open Style Lab fellow Kiley Gallant photographed by Syneos Health.
Image courtesy of Genentech.

Figure 5.7 Kangaroo pocket with magnetic zippers hoodie by Social Surge. Image courtesy of Social Surge.

Figure 5.8 Social Surge community and models. Image courtesy of Social Surge.

with disabilities and the gender nonconforming community. When creating designs, the brand analyzes the challenges PWD face every day regarding dressing with confidence and style. For example, one of the collection's unique hoodie features directly resulted from the team identifying a problem with kangaroo pockets: cell phones would often fall out when transferring out of a wheelchair. Most importantly, Social Surge product pictures demonstrate a mixed community of diverse people.

Visibility is essential to disability identity. With the mission "to resist society's desire to render [disability] invisible," the company Rebirth Garments introduces bright and bold colors for queer and disabled trans people (Cubacub 2015). Garment maker Sky Cubacub describes their lived experience of disability "of an undiagnosed stomach disability" as the launching point for their company, "I needed to wear a lot of things with much softer waistbands," says Sky (2015). Their sensory sensitivities are shared among many Americans, and the company amplified this need through stylish designs, such as the xyr neon pink crop top and bright black and white patterned shorts.

Rebirth Garments

Figure 5.9 Photo by Grace DuVal of models from left to right: Sky Cubacub of Rebirth Garments, Alice Wong of Disability Visibility Project, and Nina Litoff. All garments and accessories by Rebirth Garments.
Image courtesy of Grace DuVal and Rebirth Garments.

Most importantly, Sky takes pride in providing the option for customers to integrate various color choices. Sky recalls the frustration of finding predominantly fashion brands targeting seniors when conducting early research on fashion for PWD (Cubacub, pers. comm., 2022).

Scaling Adaptive Fashion

From bespoke clothing stores to large retail companies, the production of fashion is something every designer considers at some point during the creative process. For example, Rebirth Garments are custom designs made to order—a vertical business that also relates to many slow fashion brands. With a direct-to-consumer approach, engaged customers are essential to retail business. Kohl's attention to customers has expanded into men's and women's adaptive apparel. Kohl's is "committed to all families, aims to celebrate differences by helping consumers see themselves reflected in our brands," understanding that customers are more engaged when they can see themselves represented. The models of the way people work and focus on customer needs influence how adaptive fashion businesses can scale. Today's designers shouldn't merely be asking what designs can be created but also which economy and business structures those designs will operate in—identifying ways to scale designs that can operate with distributed manufacturing, mass customization, DIY projects, and localized networks for the twenty-first century.

Kohl's Adaptive Clothing

Patti and Ricky and Zappos Adaptative are two different examples of scaling adaptive fashion, but they both function the same—collecting adaptive design products into a single online platform. Both companies also heavily emphasize disability voices and disability-driven design. Zappos's core focus on customer needs has naturally included disability communities, which has led to successful partnerships with brands such as Sorel and UGG. In an interview with the Business Development Manager of Zappos Adaptive, Dana Zumbo, she notes the company's early involvement in adaptive design since 2014:

"The foundation and the core DNA of Zappos has always been about our customers, customer experience, and customer service. One of our senior directors in finance was going through new hire training, which includes time spent in the call center for all employees. He took a call from a customer who was returning a pair of shoes bought for her grandson, who has autism. She was having a very tough time finding shoes that were laceless—therefore, easier to get on and off. That one phone call sparked the idea behind Zappos Adaptive."

Dana Zumbo (pers.comm., 2022)

This conversation initiated the Zappos team to talk to parents with disabilities and various communities to engage in processes before launching any products on the platform. From Billy Footwear to Reebok Fit to Fit, Zappos has researched extensively into footwear needs for PWD (Dana Zumbo, pers.comm., 2022).

Conclusion 197

The platform offers AFO (ankle-foot orthoses) brace-friendly shoes with closures that wrap around the entire shoe. Another option is single shoes for people with different-sized feet or for people who have worn out only one shoe from their initial pair, such as skateboard riders.

Patti and Ricky, founded by Alex Herold, recognizes a scalable distribution of designs created with or by PWD. A medium-scale business, Alex has created a company that is sustainable in honoring its customers, creators, and quality of distribution. As the go-to adaptive fashion marketplace for adults and children with disabilities, as well as aging seniors, Patti and Ricky offer inclusive shopping. In doing so, the brand also addresses some of the technological barriers—inaccessible online shopping—a factor often missed by brands not looking at the bigger picture of disability needs.

Brands like Rebirth Garments, Zappos Adaptive, Patti and Ricky, and Social Surge are all some of the brands that demonstrate the many ways fashion businesses can scale. Designers can design for an individual, create mid-scale business models, or create large-scale ways to produce adaptive fashions. The value placed in the ways people work and the scale in which designs are produced are in question. For example, does an adapted T-shirt from a thrift store purchased for fifteen dollars hold a different value versus a mass-produced adaptive T-shirt bought from a large retailer for the same price? These comparisons in the labor, scale, and product value are more acutely illustrated in adaptive clothing.

Patti and Ricky

To better understand the complexities of adaptive fashion, designers need access to inclusive education

Figure 5.10 Billy Footwear gray shoe with wrap-around zipper featured on Zappos Adaptive.
Image courtesy of Zappos IP LLC.

Figure 5.11 Model Joshua Lhila, wearing UGG shoes for Zappos Adaptive. Image courtesy of Zappos IP LLC.

and technical skill. Many of the stories pertaining to adaptive fashion design mentioned in chapter 4 demonstrate an exchange of high levels of technical skills through collaboration. Yet these collaborative opportunities to learn about inclusive design and co-design are not readily available for all people. Public spaces are one of the ways that help people connect their own experiences to design. As mentioned in chapter 3, public spaces like museums and design spaces do serve as environments for passive contemplation as well as active learning. Finally, academia is a crucial place where adaptive fashion can make a great impact. Accessible spaces, learning opportunities, and skill sets inclusive of disability are often not introduced in classrooms. Thus, PWD can share not only their lived experiences and talents but also learn the complexity of the fashion industry, such as labor rights, recycling, material shortages, and technical skills. Additionally, educational programs provide the foundation to advance adaptive fashion by equipping people with knowledge on disability, intersectionality, and inclusive design. Some of the world's pressing challenges, like sustainability, are interdisciplinary and, therefore, must be practiced in collaborative ways. Adaptive fashion is no different. If more design programs and fashion courses included disability needs as part of a collaborative design process, this could increase disability visibility while creatively solving the fashion industry's pressing challenges. Innovative solutions and designs can be developed as a result of interdisciplinary teams. From beauty products to wearable technologies, disability provides a creative lens to the future of design possibilities.

Fashion has historically disregarded the body, writes Entwistle, who advocates for the centrality of dress to focus on people's identity, gender, and sexuality (Entwistle 2000). Therefore, it is imperative to create educational opportunities that advance fashion practice to be more inclusive and diverse, as depicted in *Fashion Education: The Systemic Revolution* (Barry and Christel 2022). With more collaborations, an exchange of disability and design knowledge could become present everywhere in education, business, and, of course, fashion.

References

Barry, Ben and Deborah A. Christel. 2022. *Fashion Education: The Systemic Revolution*. Bristol, England: Intellect.

Cabigiosu, Anna. 2020. "An Overview of the Luxury Fashion Industry." In *Digitalization in the Luxury Fashion Industry*. Cham: Palgrave Macmillan. doi: 10.1007/978-3-030-48810-9_2 (accessed August 14, 2023).

Chromat. 2021. "Chromat x Tourmaline SS22." Blog. Last modified September 15, 2021. https://chromat.co/blogs/news/ss22 (accessed August 14, 2023).

Criales-Unzueta, José. 2022. "A *Double Take* on Adaptive Fashion at NYFW, from Open Style Lab." *Vogue*, September 8. https://www.vogue.com/article/adaptive-fashion-show-at-nyfw (accessed September 16, 2022).

Cubacub, Sky. 2015. "Radical Visibility: A Queercrip Dress Reform Movement Manifesto." https://rebirthgarments.com/radical-visibility-zine (accessed August 14, 2023).

Dunn, Michael. 2022. Genentech Adaptive Fashion. edited by Grace Jun.

Entwistle, Joanne. 2000. *The Fashioned Body: Fashion, Dress and Modern Social Theory*. Cambridge: Polity Press.

Farra, Em. 2021. "This Is What's Missing in Fashion's Inclusivity Movement." *Vogue*, August 27 https://www.vogue.com/article/whats-missing-in-inclusivity-movement-adaptive-fashion-disabled-community (accessed August 14, 2023).

Kaufman, Jonathan. 2022. "Mindset Matters: The Rise Of Adaptive Fashion Awakens More Than Just A Business Opportunity." *Forbes*. https://www.forbes.com/sites/jonathankaufman/2022/09/23/mindset-matters-the-rise-of-adaptive-fashion-awakens-more-than-just-a-business-opportunity/?sh=6c6d3ab719e8 (accessed September 24, 2022).

Rawlins, Aimee. 2021. "This Disability Activist Is Pushing Starbucks, Gucci, and Others to See the Beauty of Accessibility." *Fast Company*. https://www.fastcompany.com/90652278/most-creative-people-2021-sinead-burke (accessed August 14, 2023).

Afterword

Steven Faerm, Associate Professor of Fashion at Parsons School of Design, The New School, USA

There has never been a greater urgency for the design industries to understand and address the interconnectedness of our complex social systems. It is critical for design practices, projects, and the exchange of ideas to span disciplinary boundaries and be able to inspire collaboration. The increasing complexities of these systems—and the inherent potential for design innovation when traditionally "siloed" practitioners interact together—require designers to gain advanced contextual intelligence so that the attendant industries will flourish rather than languish in the coming decades. *Fashion, Disability, and Co-design* plays a pivotal role in promoting this integrated world for innovation, one in which design transcends conventional forms of material and aesthetic value by delivering critical social value and well-being for all.

In the practice of fashion design, this necessitates a shift away from the mere drafting and styling of apparel ("product creation") to designing fashion products and systems that aim to create a better society for humanity going forward ("process creation") (Muratovski 2010). Adaptive fashion advances this shift through its inherent character and associated design processes that situate designers and wearers/consumers in a shared discourse throughout a design's development. From nascent, speculative research and prototyping to final product outcomes that strategically meet end users' needs, adaptive fashion creates a more equitable, just, and comfortable life for everyone. Thus, adaptive fashion offers substantial value to both industry and society alike due to its ability to promote economically viable design practices and support "good" causes.

The opportunities for adaptive fashion to achieve deep, positive social impact and sustained fiscal success in the marketplace are great, given the scope of fashion today. The fashion industry's rapid, unprecedented growth and seismic shifts in consumer behavior that began in the early 2000s have led design to permeate nearly every facet of our daily lives. This is due to several factors, including: the ubiquity of the internet, which has promoted higher standards of choice and aestheticism among consumers; the increase of marketing campaigns that pushed conspicuous consumption; growing affluence; and the emergent technologies that facilitate hyper-accelerated production output (Faerm 2021). Today, the fashion industry is a behemoth that is seemingly inexhaustive, producing more and more items, faster and faster than ever before. In the span of just three decades, it has grown from a $500 billion industry to its present $2.4 trillion a year global enterprise (Thomas 2019).

Yet, not everyone in the world has equal access to or the same needs as mainstream fashion. For the estimated 1.3 billion people (1 in 6 worldwide) who experience significant disability, apparel must strategically respond to and better support their specific needs and lifestyles (World Health Organization 2022). Such needs and attendant lifestyles that require adaptive fashion will likely increase, particularly among the global population aged sixty-five years or over, which is expected to double from 703 million (1 in 11) to 1.5 billion persons (1 in 6) between 2019 and 2050, respectively (United Nations 2019). Correspondingly, projections for profitability in adaptive fashion are high: In 2021, the global adaptive clothing market was valued at approximately $334.5 billion, and it is estimated to grow at nearly 5 percent CAGR in the forecast period 2022–2030 (Coherent Market Insights 2023). It, therefore, behooves fashion design companies—and fashion design educational and training programs—to educate their designers and students in adaptive fashion practice so they can enter and thrive in this highly lucrative sector of apparel design.

As the cornerstone of every society, education plays a critical role in bettering the world by training students to generate thoughtful proposals and subsequent solutions for social innovation and people's well-being. For fashion design education, this generates an existential question: How can fashion design education play a more meaningful, impactful role in society? Rather than continuing its legacy of teaching students how to produce more and more piles of pretty things, fashion education (and its educators) must come to better understand the wider potential, opportunities, and obligations it has to society (Chochinov 2007). Fashion design education must move to become issues-driven; this constitutes a dramatic shift from the previously dominant paradigm of a creative practice exclusively centered on commerce to one that is focused squarely on improving society.

In this advanced academic model, students are taught to ask different and more macro-level questions about the fashion academy and the industry—namely, questions about the ways they can adopt more altruistic roles as fashion designers, how their work can better serve consumers' needs, and how they can, collectively, support all members of society. Curricula must encourage students' openness to integrate new insights into the design process itself so that students may gain design dexterity and become nimble in working across different design spaces—both literally and figuratively. By prioritizing issues-driven (adaptive) fashion design practice—a practice that considers not only the problems but also the context of the problems affecting the world at large—the academy will produce more multidisciplinary fashion designers, ones who are better equipped to address issues relating to social responsibility, design activism, sustainability, ethnography, and systems thinking. Thus, the fashion design academy advocates for designers to build better worlds as much as it prioritizes design utility and aesthetics.

The resultant porousness between formerly siloed design disciplines and the increasing overlap with alternative academic fields that include the social sciences, body physiology, ergonomics, and engineering focused on the needs of PWD will directly prepare students to engage in adaptive fashion design methodologies. Moreover, students' engagement in adaptive fashion practices will equip them with the most desired design skills of the future, which, according to a recent survey of over 9,000 design professionals, include: designing for disabilities; understanding the ethical role of designers; cultivating empathy in design; adaptability (to societal change); cross-functional/multidisciplinary skills; and systems thinking (American Institute of Graphic Arts & Google 2019). By incorporating these curricula, fashion design education will remain relevant to both future job markets and holistic design practices (Faerm 2023). In turn, these designers and schools will foment increased awareness of and support for issues relating to social justice, community, and design innovation around the world.

Implementing pedagogy in adaptive fashion design is critical if we are to evolve the global fashion system's emphasis on commercial value by incorporating social innovation and increasing the societal value of design products. After all, educators instill in students the habits and ways of being in design practice that graduates later embody and champion in the global fashion system. Throughout all course syllabi, adaptive fashion design processes—such as performing ethnographic research, incorporating human-centered design principles, conducting case studies, and analyzing users' behavioral patterns and ergonomics—*must* be taught. In doing so, the conventional designer–user relationship is radically transformed into a more authentic, meaningful, purposeful, and socially responsible collaboration. Students will graduate as citizen designers who have a greater capacity to understand the implications that their fashion designs can have in society. The very role of the fashion designer is propelled forward from the staid, hierarchical position of the designer-as-auteur, whose personal proclivities and dictates are blindly followed by consumers, to the designer-as-altruist, someone who, through their design processes rooted in human-centered design, offers compelling products that provide societal value.

Through this advancement in fashion education, graduates will also be better prepared to engage with Generation Z consumers (individuals born between the mid-1990s and early 2010s), who currently make up nearly 40 percent of consumers worldwide (Fromm 2022). They have been raised in a world increasingly fraught with global political tension, societal violence, environmental crises, and social instability. Accordingly, Gen Z consumers expect more from the brands they patronize from a social perspective. They prefer products closely associated with the advancement of the collective good. In short, Gen Z shoppers use consumption as a means to express their core values, which are firmly rooted in the concept of social justice. For example, recent studies reveal that 94 percent of Gen Z consumers believe companies have a responsibility to address social and environmental issues, and 92 percent would switch brands to one associated with a good cause, given similar price and quality (Cone Communications 2017). Another survey found that 66 percent of global respondents said they were willing to pay more for products and services from companies that are committed to positive social change and environmental impact (Thomas 2019). When purchasing products, consumers today rank social and environmental responsibility as two of the top five factors they consider before buying a product (Thomas 2019).

Moving forward, consumers are looking beyond products and services by closely examining the respective brand, its mission and purpose, and if/how the company behind the brand and its offerings positively contribute to society. Thus, a brand's practices—and not merely the physical, aesthetically pleasing products they offer—factor heavily into the purchasing decisions and brand loyalties of this new generation of consumers. In the context of fashion design, while a garment and associated brand must deliver tangible value, fashion must also provide consumers with advanced forms of the aforementioned intangible social values. For many leading fashion brands, including Tommy Hilfiger, JCPenney, Zappos, and Kohl's, the offering of adaptive fashion lines has allowed them to become more issues-driven, serve the greater good, and connect more meaningfully with their audiences.

At its core, adaptive fashion is about helping cultivate a more inclusive society. It helps break down barriers and promotes social equity by making it easier for PWD to participate in mainstream fashion and, in doing so, can help increase the visibility and representation of these individuals. The creation of this social equity rests in part on the global fashion industry, a global powerhouse that must consider more deeply its obligations to society—namely by promoting and engaging more widely in adaptive fashion design practice.

Fashion design education must evolve the ways it prepares students so that they emerge from the academy and become professionals who will lead this global industry. With an education that places importance on all of the above-mentioned critical issues, graduates can become citizen designers, a role that places far greater responsibility on the shoulders of fashion designers moving forward, with responsibilities to the marketplace that go above and beyond simply producing a garment/product and generating profit. Coursework in adaptive fashion design will furnish students with levels of understanding of human-centered design, collaboration, multidisciplinary work, strategic planning, systems thinking, problem-solving, ergonomics, and the role of empathy in creative practice far beyond what has been seen in any preceding generation. All of these elements are significantly increasing in demand across all sectors of the fashion industry.

References

American Institute of Graphic Arts & Google. 2019. "AIGA 2019 Design Census." https://designcensus.org/ (accessed August 14, 2023).

Chochinov, Allan. 2007. "1000 words: A Manifesto for Sustainability in Design." *Core 77*, April 6. https://www.core77.com/posts/40586/1000-Words-A-Manifesto-for-Sustainability-in-Design (accessed August 14, 2023).

Coherent Market Insights. 2023. "Adaptive Clothing Market Analysis." Market Insight. https://www.coherentmarketinsights.com/market-insight/adaptive-clothing-market-2294 (accessed August 14, 2023).

Cone Communications. 2017. "2017 Cone Gen Z CSR Study: How to Speak Z." *Porter Novelli.* https://www.porternovelli.com/wp-content/uploads/2021/01/17_2017-Cone-Gen-Z-CSR-Study-How-to-Speak-Z.pdf (accessed August 14, 2023).

Faerm, Steven. 2021. "Evolving 'Places': The Paradigmatic Shift in the Role of the Fashion Designer." *Fashion, Style & Popular Culture* 8 (4): 399–417. doi:10.1386/fspc_00099_1 (accessed August 14, 2023).

Faerm, Steven. 2023. *Introduction to Design Education: Theory, Research, and Practical Applications for Educators.* Oxfordshire, England: Routledge.

Fromm, Jeff. 2022. "As Gen Z's Buying Power Grows, Businesses Must Adapt Their Marketing." *Forbes*, July 20. https://www.forbes.com/sites/jefffromm/2022/07/20/as-gen-zs-buying-power-grows-businesses-must-adapt-their-marketing/?sh=781579d82533 (accessed August 14, 2023).

Muratovski, Gjoko. 2010. "Design and Design Research: The Conflict between the Principles in Design Education and Practices in Industry." *Design Principles and Practices: An International Journal—Annual Review* 4 (2): 377–386. doi:10.18848/1833-1874/CGP/v04i02/37871 (accessed August 14, 2023).

Thomas, Dana. 2019. *Fashionopolis: The Price of Fast Fashion & the Future of Clothes*. London: Penguin Books.

United Nations. 2019. "World Population Prospects 2019: Highlights." Publications. https://population.un.org/wpp/Publications/Files/WPP2019_Highlights.pdf (accessed August 14, 2023).

World Health Organization. 2022. "Disability." Newsroom. Last modified March 7, 2023. https://www.who.int/news-room/fact-sheets/detail/disability-and-health (accessed August 14, 2023).

Acknowledgments

I cannot pick a single moment where I had begun to study this topic, but my interests grew because of collective experiences in my life: my personal experiences with disability, designing accessibility features for smartphones, growing a non-profit organization, constructing clothing with seamstresses in New York's garment district, and finally learning about disability from the many scholars, activists, and disabled leaders before me. These experiences have shaped my research and practice which is centered on human-experiences. I am first grateful to my supportive family for providing opportunities to put my "work" into context. To my parents, Myungsook Jun and Jae Whun Jun, who championed my choices in the arts, helped me discover creative solutions, and instilled within me a fierce confidence to pursue a career in design. I know design is powerful and meaningful because of them. To my thoughtful and caring husband Dr. Gregory D. Lee, who spent several nights reading and listening to my revisions. I could not have written this book without his patience and honest feedback. My love to Joung Hee Lee, George Lee, Jean Pak, Kangbin, Dobin, and of couse Lily.

Writing this book has provided me with a way to organize and speak about the importance of multidisciplinary and interdisciplinary practice. I could not have accomplished this without all the discussions and encouragement of peers who understood the value of collaboration, especially during my time as an assistant professor of fashion at Parsons. I have learned much about fashion and technology by being surrounded by such talented faculty. I would like to first thank all my colleagues at AMT and the School of Fashion. I am grateful to the people who helped me move from rough ideas to more fully developed arguments, especially Sven and Katherine. To Katherine Moriwaki and Audrey Sutton for looking over my early drafts and outlines, while providing insightful discussions on how to frame disability and design. To Sven Travis for his invaluable insights into the relationship between design and technology. My gratitude to former dean, Burak Cakmak for championing my first courses in inclusive fashion design. Finally, to my former colleagues Fiona Dieffenbacher, Marie Genevieve Cyr, Hazel Clark, and Timo Rissanen for their support in welcoming my research to the school of fashion.

I am also grateful to the following people who have pushed me toward this project prior to publishing this book. To Pamela Horn at the Cooper Hewitt, who took the time one long afternoon to help organize my outlined thoughts. Dr. Smita Rao for broadening my research approach through the lens of physical therapy. Lea Minha Yoon for connecting me to Matthew Cavnar for my first discussion on getting an agent before I knew what writing a book truly entailed. Heather Lindell Tally for her moral support and belief in my work. My deepest appreciation to Victoria Cundiff, for reminding me that books have copyright laws, therefore inspiring me to record every disability and design experience I could recall.

I am thankful to various organizations, institutions, and universities that offered the opportunity to work through ideas contained here. They include the Graphic Design Area at the University of Georgia (UGA), Parsons School of Design, the National Endowment for the Arts, and Open Style Lab. I appreciate permission from the RUSK Rehabilitation at NYU Langone Health, Zappos Adaptive, RogLab, the Levi Strauss & Co. Archives, and Heartist Team at the Samsung C&T Corporation to use images and interviews found in Chapters 1 and 2. To the brands and individuals who took time to speak with me and provided usage of images throughout this book: Rebirth Garments, Patti & Ricky, Chromat, NOT, SMA My Way program, Kohl's, Social Surge, Feeldom Life, JuneAdaptive, Chamiah Dewey Fashion, and Werable. I am grateful to Professor Julie Spivey for encouraging me to pursue time off to finish writing, the UGA Arts Fellowship Award for granting me the time to write this book and Ben Britton, who generously offered his time after one Crit Club meeting over cookies. A big thank you to the graphic design area professors and scholars for looking over my last-minute cover designs: Moon Jung Jang, Annika Kappenstein, and Erin Moore. Finally, to my two CURO research assistant students at UGA: Ellery M. Payne, for creating the fashion illustrations featured in the tutorial section of this book; and Rachel Gahyun Park for organizing the photo archive database.

To the people who have helped me build this book. Thank you to Francisca Ovalle for her attention to detail and helping me refine my ideas in a way that made this book more compelling. My gratitude to Sara Hendren, for providing the foreword to this book that thoughtfully captures the complexities of holistic design. Sara's attention to look over and refining the second chapter of this book was invaluable. To Steven Faerm, for helping me revise my first chapter with such late notice and writing the afterword for this book that connects adaptive design to the future of design education. To Yasmin Keats in editing the adaptive fashion brand resource list and Patricia Alves-Oliveria for revising parts of Chapter 3. To Alex Tosti for beautifully photographing my early work on adaptive jacket designs with Dorothy Jones. My deepest appreciation to Dorothy, who has taught me about the needs of women living with breast cancer.

My thanks to the academic scholars who shared discussions with me during the worst times of the academic year when most wish not to be bothered – Bess Williamson, Jessica Glasscock, and Ellen Fowles. To Professor Jongbae Kim from Yonsei University for discussing with me wearable factors inclusive of spinal cord injury. To Cara McCarthy for providing her insights on disability representation in museum spaces, Staci Chan for her in-depth research in assistive tools as an occupational therapist, and Dr. Jo Gooding for her specialized disability fashion research. To all my past students over the years, whose questions and insights have helped me develop my thinking and approach to design.

My deepest gratitude to Christina Mallon, Michael Tranquilli, Angela Domsitz Jabara, and Dr. Jeanne Tan for the insightful interviews, enriching the scope of fashion design. A special thank you to all the people who I have met through Open Style Lab and Parsons School of Design for placing trust in me to retell their lived experiences featured in Chapter 4: Douglas Balder, Emily Ladau, Jim Wice, Justin Moy, April Coughlin, Quemuel Arroyo, Silvo Mehle, Josh Gilnsky, Colleen Roche, Eliza Mury, Aimee Mury, Christina Mallon, Rachel Handler, Mia Piaget, Ada Stewart, and Roxine Gaussite. I want to also thank Ashley Romano and Cathy Diamond from CareRite Center, Riverside Rehabilitation for their patience and efforts in helping to gather materials to narrate some of the outcomes of an adaptive fashion collaboration in 2018 between Riverside Rehab and Open Style Lab. To Inanc Eray and Pinar Guvenc for our many collaborations with SOUR. My deepest gratitude to Dr. Grace Teo, Lea Yoon, and Alice Tin for entrusting me Open Style Lab, which celebrates its tenth-year anniversary coincidently with the same publishing year of this book.

My heartfelt thanks to Georgia Kennedy, editor at Bloomsbury Publishing, who took a chance on the Korean American girl living under the cupboard who didn't know she had magical powers to write over 60,000 words. To Rosie Best for carefully reviewing my ideas after peer review. I must also acknowledge the insightful critiques offered by the external reviewers which has helped me refine my manuscript. Finally, my gratitude to Dave Wright for his patience with my many pink styled revisions.

Further Resources

Below is a list of some of the many companies and brands at the intersection of fashion and disability. Language is reflective of that featured on the websites.

Adaptive Companies and Brands in the United States

Abilitee
Abilitee is an inclusive clothing brand with the mission to grow the conversation about disability, accessibility, and inclusion—through the language of fashion. https://abilitee.com/ [Inclusive clothing and accessories]

ABL Denim
ABL Denim is a denim collection with adaptive features, making them easier to get on and off. Designed for men, women, and children with disabilities. https://abldenim.com [Adaptive clothing]

Adaptations by Adrian
Adaptive by Adrian customizes adaptive clothing for special needs. https://www.adaptationsbyadrian.com/Default.asp [Customized adaptive clothing]

Adaptive Clothing Showroom (Benefit Wear / Innova)
Adaptive Clothing Showroom provides clothing and accessories for men, women, and children in need of regular or adaptive clothing to help ease the task of independent or aided dressing. Their adaptive clothing has been designed especially for those who are bedridden, wheelchair users, or with some limited mobility. https://adaptiveclothingshowroom.com [Adaptive clothing]

AllHeart
AllHeart is an online supplier of medical apparel, footwear, accessories, and personal diagnostic equipment. https://www.allheart.com/adaptive-clothing-women [Marketplace for adaptive clothing, footwear, and medical wear]

Alter Your Ego
Alter Your Ego designs clothing for those who are wheelchair-bound or otherwise need special assistance. https://alterurego.co [Adaptive clothing]

Ankhgear
Ankhgear is a fun-loving designer who teamed up to make technology work better for everyone. https://ankhgear.com/ [Adaptive clothing and closures]

Appaman Adaptive Kids
Appaman retains its commitment to unparalleled customer service, garment quality, and, above all else, the whimsical spirit of childhood. https://www.appaman.com/search?type=product&q=adaptive [Adaptive kids clothing]

Befree
Befree's zipOns® is a brand focusing on the challenge of dressing for those with limited mobility. They have an elastic waist and zippers that span from waist to hem. They unzip completely on both sides, can be taken on and off without going up through the legs, and can be widened from the bottom as needed. https://befreeco.com [Adaptive pants]

BILLY Footwear
BILLY Footwear provides shoes that incorporate universal design. Utilizing FlipTop technology, we have revolutionized how footwear is put on and taken off. https://billyfootwear.com [Universal footwear]

Buck and Buck
Buck and Buck is an adaptive clothing design and has been manufacturing for over forty years. B&B is a source of quality clothing and personalized customer service for caregivers ranging from family members, nursing facility staff, and guardians. https://www.buckandbuck.com/ [Adaptive clothing and footwear]

CathWear
CathWear was founded with the mission to improve the quality of life for individuals who have undergone medical treatment by restoring their dignity and privacy during treatment and recovery. https://cathwear.com/ [Medical underwear for catheter users]

CBO Baby

CBO Baby is an adaptive clothing brand for toddlers and children who live with cumbersome medical equipment, as well as those children who need the comfort and convenience of a bodysuit past babyhood. https://www.cbobaby.com [Adaptive clothing brand for children]

Clothes for Seniors

Clothes for Seniors supplies traditional and adaptive clothing and footwear to nursing home residents and senior home healthcare individuals. https://www.clothesforseniors.com
[Adaptive footwear and clothing suppliers]

Cure8able*

Cure8able is a styling service for people with disabilities guided by the Disability Fashion Styling System created by Cur8able's founder, Stephanie Thomas. https://cur8able.com/
[Styling consultancy for people with disabilities]

Dunnes Stores

Dunnes Stores launched their Additional Needs Clothing Range in 2016 for ages 2–14 years old. https://www.dunnesstores.com/c/kids/highlights/additional-needs-clothing [Dunnes' additional needs clothing range for kids]

è Ispirante—Creative Adaptive Clothing

è Ispirante™ is a creative Adaptive Clothing Line that features skirts in beautiful patterns and intricate designs with hook and loop closures. This line blends function and fashion together for adults and children to allow the wearer to feel confident and empowered. https://www.6pm.com/ispirante-creative-adaptive-clothing [Adaptive clothing]

Easy Access Clothing

Easy Access Clothing produces high-quality, all USA-made adaptable clothing while offering an extremely high level of comfort. https://easyaccessclothing.com [Adaptive clothing]

Eone

Eone created a sleek, modern watch that more people can use—and in more ways. Designed for touch when you can't easily use sight: during a meeting, in a movie theater, or due to a vision impairment. https://www.eone-time.com/ [Design-forward accessible watches]

Friendly Shoes

Friendly Shoes solve more types of footwear challenges than any other shoe technology by making fitted shoes simpler and easier to put on, and more enjoyable to wear. https://friendlyshoes.com/aff/lifezest/ [Inclusive footwear]

HarperSage

HarperSage is an ethical lifestyle brand that empowers multi-faceted women through both community and style, creating thoughtful, stylish, comfortable fashion staples. https://harpersage.com/search?q=adaptive
[Lifestyle brand with adaptive pants]

Hatchbacks

Hatchbacks footwear is exclusively designed for use over AFO, DAFO, and other custom inserts. https://www.hatchbacksfootwear.com [Adaptive footwear]

Independence Day Clothing/ID

Independence Day Clothing/ID is a women-and-minority-run company drawing from the resources of the fashion, design, special education, media, and finance sectors of New York City. https://www.independencedayclothing.com [Adaptive clothing]

IZ Adaptive

IZ Adaptive is a brand that designs clothing for people with limited mobility, seated or standing. Their unique collection is crafted with style, fit, and purpose. https://izadaptive.com [Adaptive clothing]

Izzy Wheels

Izzy Wheels provide a range of stylish wheel covers for wheelchairs—with the motto, "If you can't stand up, stand out." https://www.izzywheels.com/
[Stylish wheel covers for wheelchairs]

Jansport: The Adaptive Collection

The Adaptive Collection brings together the iconic style of JanSport with the function of device-specific packs. The collection features JanSport Adaptive Backpack and Adaptive Crossbody. https://www.jansport.com/collections/adaptive-collection.html [Adaptive accessories]

Jumping Beans

Jumping Beans is a collection of bright, fun, and trendy casual clothing from multinational brands for young children ranging from newborn to 14 years of age. https://www.kohls.com/catalog/jumping-beans.jsp?CN=Brand:Jumping%20Beans [Adaptive clothing]

Further Resources

Kinetic Balance
Kinetic Balance designs chair riding apparel that makes going outdoors comfortable in a stylish way. They excel in comfort, fit, performance, and looks. https://www.kinetic-balance.com [Adaptive clothing]

Kohl's Adaptive
Kohl's Adaptive offers adaptive apparel for children and adults, ensuring everyone can feel comfortable and look their best every single day. https://www.kohls.com/catalog/adaptive.jsp?CN=Feature:Adaptive
[Adaptive clothing marketplace within Kohl's]

Koolway Sports
Koolway Sports designs outerwear for people in motion-enabling them to achieve their maximum level. https://koolwaysports.com
[Custom-made clothing for individuals with disabilities]

Kozie Clothes
Kozie Clothes provides deep pressure sensory children's clothing, weighted and compression clothing for infants to youth, medical clothing for infants and small children, weighted blankets, Lap pads, and children's sensory play throughout the Pottstown area, United States, and Canada https://www.kozieclothes.com
[Sensory-friendly clothing and products]

Lands' End Adaptive
Lands' End is a global multi-channel retailer designing and selling classically styled apparel, swimwear, and outerwear for Women, Men, and Kids, plus a complete line of home products, luggage, and seasonal gifts. https://www.landsend.com/adaptive-clothing-kids-school-uniforms-sensory-friendly/ [Adaptive clothing line]

Magna Ready
Magna Ready is a patented technology of a magnetized dress shirt created for people with limited mobility. https://magnaready.com [Inclusive shirts with magnets]

Miga Swimwear
MIGA Swimwear surveyed over 400 women to find their perfect styles, resulting in functional and comfortable swimwear that inspires confidence. All of their suits are made ethically in the United States and use only sustainable fabrics. https://migaswimwear.com
[Inclusive and sustainable swimwear]

Myself Belts
Myself Belts designed a belt that children can fasten themselves. Perfect for potty training, school uniforms, or cute accessorizing, and helpful for adult hand dexterity challenges. https://www.myselfbelts.com
[Easy fasten belts]

Nike FlyEase
Nike FlyEase lets you enjoy sport no matter your ability with technology developed from insights from the disability community. https://www.nike.com/flyease
[Nike footwear with inclusive features]

Ovidis
Ovidis offers comfortable, high-quality easy wear adaptive clothing for seniors, the elderly, and disabled persons, helping individuals get dressed more easily and gain autonomy. https://ovidis.com [Adaptive clothing]

Patti and Ricky*
Patti and Ricky is the Adaptive Fashion Marketplace™ for adults and children with disabilities, chronic conditions, patients, seniors, and caregivers. https://www.pattiandricky.com [Adaptive fashion marketplace]

PreventaWear
PreventaWear make special needs clothing for children and adults, including onesies, to cater to those with incontinence and their caregivers with adaptive clothing. https://preventawear.com [Special needs clothing]

Propét Adaptive
Propét is a rapidly growing shoe company dedicated to making the world's best walking shoes. Propét shoes are better walking shoes because they're designed and manufactured to be your most comfortable walking experience ever. https://www.6pm.com/propet-adaptive [Adaptive footwear]

Reboundwear
Reboundwear designs and manufactures easy dressing apparel for post-surgery, physical therapy, elderly, those with disability, and anyone with limited mobility. https://www.reboundwear.com [Adaptive clothing]

Reebok Fit to Fit Shoes
Reebok's Fit to Fit collection is a design built upon Reebok's iconic design heritage and silhouettes. The collection's goal is to enhance quality of life for everyone by providing functional products that don't compromise on style or performance. Each model within the collection

offers enhanced features to help people with disabilities gain more independence in whatever they do. https://www.zappos.com/e/adaptive/reebok-fit-to-fit-collection [Reebok's adaptive footwear collection]

Sensory Smart Clothing Co.
Sensory Smart Clothing Co. is a line of kidswear made specifically with sensory sensitive children in mind. Every piece is created with special features like no tags or outside seams and made from ultrasoft fabrics to ensure it is comfortable for all children, even those with tactile sensitivities. https://sensorysmartclothing.com [Sensory-friendly clothing]

Seven7 Adaptive
Designed in America, Seven7 able jeans adaptive denim collection offers simple solutions to dressing so that everyone's personality can shine through their wardrobe. https://seven7jeans.com/shop/adaptive-men [Seven's adaptive clothing line]

Silverts
Silverts is an adaptive clothing and footwear brand that is designed to make the lives of PWD and their caregivers lives easier. Silverts provides simplified dressing for empowered living. https://www.silverts.com/ [Adaptive apparel]

Slick Chicks®
Slick Chicks® features a hook and eye fastener system that lets any woman seamlessly transition in and out of her panties. https://slickchicksonline.com [Adaptive clothing]

Smart Adaptive Clothing
Smart Adaptive Clothing is an adaptive clothing brand universally designing for someone living with a disability, chronic illness, seniors, and anyone struggling with dressing. Providing high-quality, fashionable garments that are easy to wear, washable, and individual. https://smartadaptiveclothing.com/ [Adaptive clothing]

SmartKnit
SmartKnit Seamless Socks are made using a knitting process much like how a cocoon is spun, starting at one point and spinning upwards, thus completely eliminating irritating seams and the risks associated with them. SmartKnit socks feel like a second skin and provide comfort and relief for those who have diabetes, arthritis, or sensitive feet. https://smartknit.com [Seamless socks]

So Yes
So Yes is a company and a brand that wants to make people shine with an adapted clothing offer. So Yes goes to great lengths to make people with and without physical limitations feel great by designing, producing, and selling specific, beautiful, high-quality, and affordable clothing. https://so-yes.com/en/home-so-yes-adaptive-clothing/ [Adaptive clothing and footwear brand and supplier]

Social Surge
Social Surge is a universally designed apparel brand that is rooted in the core belief that all people are more alike than different. https://www.socialsurgeofficial.com [Universal apparel]

Special Kids Company
Special Kids Company offers a range of adaptive clothing and accessories designed for children with special needs. They are passionate about creating sensory clothing and accessories, such as wheelchair covers and chewies, that can enhance a child's social skills by addressing their unique needs without sacrificing their personal style and self-expression. https://specialkids.company [Adaptive clothing and accessories for children]

Spoonie Threads
Spoonie Threads aspires to create happiness, spread optimism, and simplify life. Featuring adaptive clothing and accessories for those battling cancer, disabilities, and chronic illnesses. https://spooniethreads.com/ [Adaptive apparel and accessories brand]

Steve Madden
Steve Madden's Adaptive shoe collection is designed to help kids learn to put on footwear by themselves, no matter their level of ability. Learning to dress independently helps kids feel comfortable, confident, and ready to take on the world. Their collection of adaptive kids' shoes features a wide range of easy-on styles. https://www.stevemadden.com/collections/kids-shoes-adaptive [Steve Madden's kids adaptive footwear collection]

Stitches Medical
Stitches Medical was founded to answer a simple patient question: What should I wear? Founded by practitioners and patients, Stitches Medical recognized the need for apparel that is both functional and fashionable for patients with a variety of medical conditions. https://stitchesmedical.com [Adaptive apparel]

Target Adaptive
Target's adaptive collection is designed to be functional, sensory-friendly, and easier-to-access to fit the needs of babies and kids. https://www.target.com/c/women-s-adaptive-clothing/-/N-r23zu [Target's adaptive clothing]

Tommy Hilfiger
Tommy Hilfiger Adaptive was created with one goal—to deliver classic, American cool style with innovative design twists that make getting dressed easier for the entire family. https://usa.tommy.com/en/tommy-adaptive [Tommy Hilfiger's adaptive clothing]

Tony and Ava
Tony and Ava is a brand that believes that every child should have the right to his or her own pair of underwear, whether it's regular or adaptive. Tony and Ava design so that every child can be comfortable and have assured freedom of mobility in breathable, leak-controlled underwear. https://tonyava.com/
[Regular and adaptive underwear for kids]

Two Blind Brothers
Two Blind Brothers is a luxury clothing line on a mission to cure blindness. https://twoblindbrothers.com/ [Marketplace where customers shop 'blind']

UGG Universal
UGG Universal is an inclusive capsule, updating timeless silhouettes with specialized modifications for easy entry. https://www.ugg.com/st%20ory?id=ugg-universal [UGG's Universal footwear line]

Uniquely Regal
Uniquely Regal is a sensory-friendly clothing brand—where function meets fashion for special needs children and teens. https://www.uniquelyregalkids.com [Sensory-friendly clothing brand]

UNYQ
UNYQ provides custom-made, stylish prosthetic wears for amputees. By reimagining something that used to be a source of frustration into a means for self-expression with the help of biometrics, 3D imaging, and 3D printing. https://unyq.com/ [Custom-made stylish prosthetics]

Vans
Van's sensory-inclusive footwear designs are part of the Autism Awareness Collection. With this project, Vans celebrates the unique aspects of all people. Designed specifically with ASD in mind, the ultra-comfortable footwear collection is offered with sensory-inclusive elements, including a calming color palette and design features that focus on the senses of touch, sign, and sound. https://www.vans.com/en-us/article-detail/autism-awareness [Van's sensory-inclusive footwear line]

Wardrobe Wagon
Wardrobe Wagon is an adaptive and traditional wearing apparel brand. https://www.wardrobewagon.com [Adaptive brand for clothing, footwear, and medical wear]

YoRo Naturals
Remedywear™ creates clothing for psoriasis. They are the perfect solution for treating more advanced forms of psoriasis, eczema, and more. All their garments are dry wrap friendly and protect skin from scratching. https://yoronaturals.com/collections/remedywear-adults [Clothing and accessories for sensitive skin]

Zappos Adaptive*
Zappos Adaptive is an adaptive marketplace for clothing and footwear that creates functional and fashionable products to make life easier. They carry a wide range of adaptive clothing and shoes with unique features to fit a variety of needs https://www.zappos.com/e/adaptive [Adaptive marketplace for clothing and footwear]

* Product platform or fashion stylist related

Adaptive Fashion Designers and Brands International

Able2 Wear—UK Based
Able2 Wear is a UK supplier of wheelchair clothing and adaptive clothing. They have been working closely with wheelchair users, caregivers, and professionals for over 25 years to develop a range of disabled clothing to meet specialist needs. https://www.able2wear.co.uk [Specialized wheelchair and adaptive clothing]

Adaptista—UK
Adaptista is an online marketplace for adaptive and inclusive fashion brands. https://adaptista.com/ [Marketplace for inclusive fashion]

AnnaPS—UK
AnnaPS makes comfortable, good-looking, eco-friendly clothes with integrated pockets for diabetes devices such as insulin pumps and pens, dextrose, hand units, and glucose meters. https://www.annaps.com/ [Diabetic friendly clothing]

AtoZED — Switzerland
AtoZED is a project that promotes and develops inclusive design solutions. https://atozed.ch/index.html [Ability driven design]

Bealies Adaptive Wear — UK
Bealies Adaptive Wear is an online shop selling adaptive joggers (clothing) for wheelchair users. The product is the brainchild of Caron Mcluckie, whose son suffered a spinal cord stroke in 2016. https://www.bealiesadaptivewear.co.uk [Adaptive jogger]

Chamiah Dewey Fashion — UK
Chamiah Dewey Fashion is a fashion brand for people of short stature, under 4 ft. 10 in. They pride themselves on being eco-conscious, adaptive, and timeless. https://chamiahdeweyfashion.com/ [Clothing for short stature people]

Clip-Knix — UK
Clip-Knix is an adaptive, patented, award-winning front fastening underwear for women. It can be worn sitting, standing, or lying down. https://www.clip-knix.biz/ [Adaptive underwear for women]

Constant & Zoé — France
Constant & Zoé designs practical, modern, and sparkling clothing for people with disabilities. They make dressing easier and faster both for the person with a disability and for relatives and professional caregivers. https://constantetzoe.com/pages/notre-histoire [Adaptive clothing]

Dawn Adaptive — Malaysia
Dawn Adaptive is a Malaysian adaptive clothing brand, a Social Enterprise dedicated to designing adaptive clothing that is fashionable, comfortable, and affordable for all. Their clothing is a functional solution for people with dressing difficulties or simply anyone who wants to experience a more easy dressing process. https://dawnadaptive.com/pages/about-us [Adaptive clothing]

EveryHuman — Australia
EveryHuman is a marketplace for inclusive fashion based in Australia and New Zealand. They stand for accessibility and choice, providing people of all abilities with fashion to suit their needs. https://everyhuman.com.au [Marketplace for inclusive fashion]

Ézé Plus — Canada
Mode Ézé Plus is an adaptive clothing brand. Their goal is to create adaptive garments that follow current fashion trends. Thanks to Mode Ézé Plus, people with mobility challenges or disabilities can enjoy wearing fashionable, practical, and comfortable clothes. https://www.ezeplus.com/?lang=en [Adaptive clothing]

FEELDOM Life — South Korea
A company of designers and craftspeople in South Korea working with disability communities to create higher-quality wheelchair bags. Three key products featured are the JAYU, MAX, and BUDDY bags. https://www.feeldomlife.com/ [Adaptive bags and accessories]

Free Form Style — Spain
Free Form Style is an adaptive clothing brand with the aim to allow people to be the best version of themselves while at the same time being comfortable. https://freeformstyle.com/ [Adaptive clothing and tailoring]

Friendly Shoes — UK
Friendly Shoes is a footwear brand for people of all abilities. Their shoe technology makes fitted shoes simpler, easier to put on, and more enjoyable to wear https://friendlyshoes.com/ [Footwear for all abilities]

Heartist — South Korea
Heartist is a universal design brand that reduces the inconvenience of garments for PWD and applies details optimized for disabled people for anyone to feel and enjoy fashion. https://rb.gy/fquef [Universal design brand]

inc kid — Australia
inc kid is a clothing brand that designs clothing that fits orthotics and prosthetics through big openings at the neck, sleeve, and leg for easy dressing. https://inckid.com/ [Inclusive kids clothing]

June Adaptive — Canada
June Adaptive is a marketplace for adaptive fashion for people frustrated by inaccessible buttons and zippers, recovering from medical treatments, and living with mobility challenges or disabilities. https://www.juneadaptive.com/pages/our-story [Marketplace for adaptive fashion]

Limonata Adaptive — Australia
Limonata is an adaptive clothing brand for when life gives you lemons. Their range of functional fashion helps you feel comfortable and confident—at home, in hospital, and in between. https://www.limonata.com.au [Adaptive clothing]

Optivus — UK
Optivus is the first clothing line inspired by streetwear that is specifically designed for persons who have physical limitations. https://www.optivus.uk
[Clothing brand for people with physical limitations]

Rackety — UK
Rachety creates easy-access clothing for children and adults. https://www.disabled-clothing.co.uk
[Easy-access clothing]

Roll With Style — UK
Roll With Style is a styling and fashion advice service for wheelchair users. Founded from Sandie Roberts and Natalie Care's combined passion for helping others with their years of experience and love of fashion. https://www.linkedin.com/company/roll-with-style/
[Styling consultancy for wheelchair users]

Rollitex — UK / Germany
Rollitex is a brand based in Germany that has recognized the need for highly specialized clothing for people using a wheelchair. They offer a wide variety of wheelchair pants, wheelchair jeans, and other adaptive clothing for men, women, and children alike. https://www.rollitex.co.uk/en/home/ [Clothing for people who use wheelchairs]

SAM Sensory & More — Belgium
SAM combines well-being and fashion into a label that looks good and feels even better. Children, teenagers, (young) adults, and the elderly find peace in their no-stress collection. https://www.samsensoryclothing.com/en/
[Sensory-friendly clothing and accessories]

SB Shop Diverse You — UK
SB Shop Diverse You is an online shop that curates inclusive and adaptive items. https://sbshop.co.uk/
[A brand focused on inclusive solutions through fashion]

Seated Sewing — UK
Founded by Kat, an "inclusive seamstress," Seated Sewing handcrafts luxury weighted blankets, skin protector pads, patient pamper hampers, and adaptive clothing for the disabled community. https://seatedsewing.co.uk/?page_id=12647
[Inclusive seamstress]

The Able Label — UK
The Able Label designs, makes, and retails women's adaptive clothing. All their clothes are specifically designed to make dressing easier with the perfect balance of function and fashion. https://www.theablelabel.com/ [Adaptive clothing]

Tubie Kids — UK
Tubie Kids is an adaptive clothing brand with discreet feeding tube access for children and pre-teens. Compatible with PEGS and Mic-Key Buttons. https://tubiekids.co.uk [Tube-friendly children's clothing brand]

Unhidden Clothing — UK
Unhidden Clothing is a fashion line specifically created for people with disabilities to make their lives easier and more comfortable. Everything is made to measure so everyone can look and feel their best. https://unhiddenclothing.com/pages/about-unhidden
[Universal design]

Well Cool Clothing — UK
Designed to empower, Well Cool Clothing is an innovative and practical nightwear and daywear clothing brand for hospital and homecare. https://www.wellcoolclothing.com/
[Easy-access patient clothing]

Yamato — Japan
Yamato is a kimono specialty shop with over one hundred years of history, offering new ways to enjoy Japanese garments that meet the times while handing down the tradition of kimono culture and technique. https://store.kimono-yamato.com/Page/Feature_2020SS_yukatazero.aspx [Ready- or tailor-made kimonos]

Further Reading

The following annotated reading recommendations provide more information about a topic or the content of each chapter, offering further clarification for specific words or phrases used throughout this book. For example, the term "disability" has rich implications from various studies by scholars and writers. All resources are organized in alphabetical order.

CHAPTER 1

Annett-Hitchcock, Kate. 2023. *The Intersection of Fashion and Disability: A Historical Analysis.* Bloomsbury Publishing.

This book examines the relationship between fashion and disability through notable historical figures, events, and movements, detailing the contemporary developments in clothing and fashion.

Dunn, Dana S. and Erin E. Andrews. 2015. "Person-First and Identity-First Language: Developing Psychologists' Cultural Competence Using Disability Language." *The American Psychologist* 70 (3): 255–264.

These pages provide a rationale for person-first language (e.g., people with disabilities) and identify-first language (e.g., disabled people), respectively, and are linked to models. For example, the APA (American Psychological Association) advocates the use of person-first language to reduce bias in psychological writing, while disability studies scholars promote the use of identity-first language.

Jenkins, Jo Ann. 2016. *Disrupt Aging: A Bold New Path to Living Your Best Life at Every Age*, 91–123, Public Affairs.

This book describes a new vision for living and aging by addressing age-related stereotypes. These specific pages describe the intersectionality between disability and aging with references to multi-generational homes, Alzheimer's disease, and assistive devices.

Jungnickel, Kat. 2022. *Politics of Patents (POP).* Available online: http://www.politicsofpatents.org/.

A five-year research project funded by a European Union Horizon 2020 ERC Consolidator Grant. This project aims to reimagine citizenship through clothing inventions from 1820–2020.

Ladau, Emily. 2021. *Demystifying Disability: What to Know, What to Say, and How to Be an Ally*, 32–38, Penguin Random House.

Discusses the intersectionality of disability, disability and identity, disability types, culture, and how to talk about disability. In this book, the author prefers the use of identify-first language to acknowledge disability as a part of what makes a person who they are, rather than person-first language that acknowledges that people are seen first and not just for their disability.

Linton, Simi. 1998. *"Reclamation," Claiming Disability: Knowledge and Identity*, 1–8, NYU Press.

Invites readers to reconsider the meanings given to disability and to elaborate more on their origins. These pages highlight that disability is redefined when united by a common social and political experience.

Oliver, Mike. 2013. "The Social Model of Disability: Thirty Years On." *Disability & Society* 28 (7): 1024–1026.

Discusses the influence of the social model of disability in comparison to the medical model.

Wong, Alice. 2021. *Disability Visibility: First-Person Stories from the Twenty-First Century*. Vintage Books.

Presents first-person stories and offers rich narratives on how people identify with disability.

Williamson, Bess and Elizabeth Guffey. 2020. "Introduction: Rethinking Design History Through Disability, Rethinking Disability Through Design." In *Making Disability Modern*, 1–10, Bloomsbury Publishing.

This book brings other scholars to examine how designed objects and spaces contribute to the meanings of ability and disability between the eighteenth century and 2020. The introduction particularly highlights the social role of objects and the active role designs take to reshape the material environment.

CHAPTER 2

Clarkson, John P. 2003. *Inclusive Design: Design for the Whole Population.* Springer Science & Business.

A comprehensive narrative on inclusive design and universal design methods, individual stories, and examples that include disability and aging.

Gissen, David. 2022. *The Architecture of Disability: Building Cities, and Landscapes beyond Access*. University of Minnesota Press.

This book presents a radical perspective on architecture.

Hamraie, Aimi. 2017. "Introduction." In *Critical Access Studies in Building Access*, 1–18, University of Minnesota Press.

Presents disability studies in relation to political, cultural, and social historical American events.

Hajo, Adam and Adam D. Galinsky. 2012. "Enclothed Cognition." *Journal of Experimental Social Psychology* 48 (4): 918–925

Introduces the term "enclothed cognition" to describe the systematic influence that clothes have on the wearer's psychological processes. This journal article demonstrates that the way people dress has been shown to impact their thoughts and actions and affects how others perceive them.

Hendren, Sara. 2020. *What Can a Body Do? How We Meet the Built World*, Illustrated edition. Riverhead Books.

The book explores the places where disability is present in design, such as everyday tools and environments and the built environment. The introduction chapter in particular helps readers rethink our conceptions of disability and body.

Loschek, Ingrid. 2009. *When Clothes Become Fashion*. Bloomsbury Publishing.

Includes fashion perspectives and defining terminology such as clothing.

Mitchell, SCM. 1991. "Dressing Aids." *The BMJ* 302: 167–9. doi: 10.1136/bmj.302.6769.167.

Describes how adaptive equipment and dressing aids (reacher, sock aide, button hook, dressing stick, shoehorns, clip-on ties) can assist motor deficits.

Pullin, Graham. 2018. "Super Normal Design for Extraordinary Bodies: A Design Manifesto." In *Manifestos for the Future of Critical Disability Studies*. Routledge/CRC Press.

Presents a perspective on the human body through critical disability studies.

Sanders B. Elizabeth, and Pieter J. Stappers. 2008. "Co-Creation and the New Landscapes of Design." *CoDesign* 4 (1): 5–18.

Presents co-design methods and approaches with multiple stakeholders.

Schuler, Douglas, and Aki Namioka. 1993. *Participatory Design: Principles and Practices*. Erlbaum.

Explores meaningful and productive interactions for participants engaged with the process of technology design and use. The book offers contributions from the first Participatory Design conference held in Seattle.

Thorton, Nellie. 1990. *Fashion for Disabled People*. B.T. Batsford Ltd.

Offers diverse clothing design examples and construction instructions for disabled people.

Visser, Froukje Sleeswijk, Pieter Jan Stappers, Remko van der Lugt, and Elizabeth B. N. Sanders. 2005. "Contextmapping: Experiences from Practice." *International Journal of CoCreation in Design and the Arts* 1 (2): 119–149.

Perspectives of co-design methods from occupational therapy.

Watkins Susan. M. and Lucy E. Dunne Lucy. 2015. *Functional Clothing Design: From Sportswear to Spacesuits*. Bloomsbury Publishing.

Depicts functional clothing designs created for extreme conditions that are also applicable to adaptive fashion studies.

Williamson, Beth. *Accessible America: A History of Disability and Design*, 69–95, NYU Press.

Presents the innovation of disabled people and disability advocates in design thinking with references from American civil rights issues and public life.

Wright, Natalie E. and Liz Jackson. 2022. "Functional Fashions for the Physically Handicapped: Disability and Dress in Postwar America, Dress." *The Journal of the Costume Society of America* 48: 143–162.

Demonstrates how clothing played an important role in the establishment and maintenance of the postwar American project of independence with references to Helen Cookman's work.

CHAPTER 3

Barnfield, Jo. 2012. *Pattern Making Primer*. Barron's Educational Series, Incorporated.

One of the most comprehensive fashion design books with instructions and examples of patternmaking and sewing techniques.

Liechty, Elizabeth, Judith Rasband, and Della Pottberg-Steineckert. 2009. *Fitting and Pattern Alteration: A

Multi-Method Approach to the Art of Style Selection, Fitting, and Alteration, 3rd ed. Bloomsbury Publishing.

Presents multiple methods of style selection, fitting, and alteration.

Pullin, Graham. 2009. *Design Meets Disability*. MIT Press.

Demonstrates the relationship between design and disability. From problem-solving to more playful explorations, this book showcases interviews with leading designers about specific disability design, such as glasses and prosthetic legs.

Quinn, M. Dolores and Renée Weiss-Chase. 1990. *Simplicity's Design Without Limits*. Drexel University Design Press.

From seated design to adaptive shirts, this book offers exemplary patterns and instructions on how to create clothing modifications inclusive of disability needs.

Tsukiori, Yoshiko. 2017. *Stylish Wraps Sewing Book: Ponchos, Capes, Coats and More*. Tuttle Publishing.

Provides instructions and patterns for wraps such as ponchos, capes, coats, and jackets.

Zieman, Nancy. 2008. *Pattern Fitting with Confidence*. Krause Publications

An easy approach to pattern fitting techniques such as tucking, slashing, pinching, and pivot-and-slide methods.

CHAPTER 4

Braveman, Brent and Jill J. Page. 2011. *Work: Promoting Participation & Productivity Through Occupational Therapy*. F.A. Davis Company.

Brown, Keah. 2019. *The Pretty One*. Simon & Schuster.

Crenshaw, Kimberlé. 2016. "The Urgency of Intersectionality." TED video. Available online: https://www.youtube.com/watch?v=akOe5-UsQ2o.

Hinojosa, Jim and Marie-Louise Blount. 2014. *The Texture of Life: Purposeful Activities in the Context of Occupation* 3rd ed. American Occupational Therapy Association.

Jun, Grace and Jeanne Tan. 2018. *Universal Materiality: Wearable Interaction Design and Computer Aided Process for Accessible Wearable Solutions*. The Hong Kong Polytechnic University (Exhibition Catalogue). Available online: http://hdl.handle.net/10397/78800.

A short book with exemplary case studies focused on fashion design and applied technologies such as 3D printing and illuminating textiles for aging and disability.

Microsoft Inclusive Design methodology. Available online: https://www.microsoft.com/design/inclusive/.

Pailes-Friedman, Rebeccah. 2016. *Smart Textiles for Designers: Inventing the Future of Fabrics*. Laurence King Publishing.

This book presents the increasingly technological area of smart materials and how they may be imagined as new textiles and fabrics.

Quinn, Bradley. 2010. *Textile Futures: Fashion, Design and Technology*. Bloomsbury Publishing.

A collection of textile innovations and highlights of works from key practitioners, researchers, and industry.

CHAPTER 5

Barry, Ben and Deborah A. Christel. 2023. *Fashion Education: The Systemic Revolution*. Intellect.

A collection of seventeen essays by fashion educators and students in Australia, Canada, the United States, and the UK, who explore how to transform teaching practices to foster equity, inclusion, and social justice. The book explores approaches in the fashion pedagogy and practices beyond the narrow Eurocentric canon.

Blackford, Karen, Cathy Cuthbertson, Fran Odette, and Miriam Ticoll. 1993. "Women and Disability." *Canadian Women's Studies* 13 (4).

The editors of this issue are feminists, most of whom have disabilities, and it includes articles by women from diverse backgrounds to highlight the diversity in the lives of women with disabilities. The authors raise questions that challenge traditional feminist thinking about the body, reproductive rights, and language.

Faerm, Steven. 2023. *Introduction to Design Education: Theory, Research, and Practical Applications for Educators*. Routledge.

A practical book that offers design educators a comprehensive introduction to design education and pedagogy in higher education.

Garland-Thomson, Rosemary. 1994. "Redrawing the Boundaries of Feminist Disability Studies." *Feminist Studies* 20 (3): 583–597.

This review essay argues for the recognition of feminist disability studies within feminism. The author asserts that feminist disability studies can be in the broader area of identity politics.

Glossary

The following glossary provides a broad scope of the terms used in this book. The definitions are subject to interpretation and related to their use in fashion.

ableism A social prejudice and discrimination of disabled people based on beliefs of abilities or body types that adhere to able-body standards.

accessibility The quality of being able to access, reach, or enter services, buildings, websites, and products, emphasizing barriers faced by people with disabilities and the aging population.

adaptive design A design approach that is disability-aware and provides flexible and various ways to approach a design outcome or process—for example, through customization and personalization.

adaptive fashion A wearable object, such as accessories or clothing, that follows social and cultural trends and can be adapted to the needs and wants of a person with any disability.

AT (assistive technology) Any product or system that enhances daily living, working, and learning opportunities for a person with any disability.

circuit A closed pathway in which an electric current can flow uninterrupted. For example, a circuit can be used when designing an electrical function for a wearable garment that connects to a mobile device.

conductive material Any material that can transmit heat or electricity.

co-design To participate in a design-led approach that includes end users or community members in the process of creating design solutions.

design for all An attitude and practice that shifts the attention of designing adaptations or aids toward an inclusive design practice that endeavors to include as many people as possible in the process.

DIY A method of building or modifying independently without the aid of professionals, also known as do-it-yourself.

dress form A torso-only, three-dimensional shape used for fitting clothing that is being designed or sewn.

ethnography The study of human cultures. In relevance to designers, ethnographic research provides a way to observe users and their interactions with products in real-life situations, connecting the social sciences with design practices.

functional fashion A term used to describe a garment as a means of psychological rehabilitation and its potential to address societal challenges related to disability in the United States between the mid-1950s and late 1970s. It was most recognized by the clothing line created by designer Helen Cookman in association with the CRDF (Clothing Research and Development Foundation).

inclusive design A design practice that includes end users and consumers in the process of making.

independent dressing The act or process of an individual capable of dressing alone.

interdependent dressing The act or process of an individual who dresses with the help of another, often in various sequences.

interdisciplinary design A design approach that is collaborative and process-oriented, focusing on human-centered experiences through the collaboration of members with different disciplines of study or practice.

medical model A model of disability and aging that implies people are disabled because of their own condition and can be corrected through medication, rehabilitation, surgery, or other equipment that provides a physical remedy.

occupational therapist A specialist who takes a holistic and scientific perspective or approach toward a person's ability to fulfill their daily routines and roles, often referred to as an OT.

PWD (people with disabilities) A term to describe people who live with disabilities as a resulting disharmony between the individuals and their social or physical environment rather than their condition.

physical therapist A specialist who uses a holistic approach to help restore or reduce injury or illness that impacts body movement and function.

smart material Material, such as a color-changing fabric, that can sense and react to environmental changes, such as cooling or stretching.

social inclusion A political and social act aimed to include individuals typically discriminated against for their disability, age, financial status, or ethnicity.

social model A model of disability and aging that implies people are disabled or enabled by the social or environmental context in which they function.

universal design A set of principles developed by architect and designer Ronald Mace that focus on design that is usable to all people, regardless of their age, ability, or situation.

UCD (user-centered design) A design principle, also known as human-centered design, centered on human experiences, and how people—users, customers, communities, and audiences—are at the core of all creative activities and practices.

wearables Any material or object that is worn on the body, such as eyeglasses and wristwatches.

Index

Note: Page locators in *italic* refer to figure captions.

ableism 53, 108, 109, 154
Access & Closure 51–72
accessible features in existing fashion designs 27–30
Adaptista 34
adaptive fashion 2–21: framework 48–72; growth 200–1; scaling 197–8; visibility 194–7
affordances, design 65–8
aging 11, *11*, 27, 74, 139: interactive garments 166–7, 179–84
alteration instructions 75–86
arm sling 16–17, *17*
Arroyo, Quemuel 125–32
assistive devices 3, 40–1, 108, 180: Swipe 133–6
Avisly 173–8
Avraham, Helena 119–24

bags 13, *13*, 20, *20*, 38, *38*, 40, *41*, 108–13, 129
Balder, Douglas 168–72
blanket, lap *125*, 129, 130, *130*, *131*, 132
blind and visually impaired 13, 39–40, *39*, *40*, 81, *82*
body: forms 3, 28, *29*, 32, 75, 144, *144*; measurements 38, 72, 138, 139, 156, *156*
Bosquet, Audrey 173–8
brands 3, 8, 34, 194, 197, 198, 202
Bruno, Estee *21*, 97–100
Buyukaslan Oosterom, Evrin 185–91

Chalayan, Hussain 28, *29*
Chan, Mickey 9–11, *10*, 43
Chan, Staci 125–32
children 6, 13, 148–9
Chiriboga, Camila 38, 39–40, *40*
Chromat x Tourmaline 194, *194*
circuit designs *51*, *52*, 176, *177*, *189*, 190
closures 8, *8*, 9, 66, 72–3, *73*, 139: magnetic 73, *73*, 77–8, *77*, *78*, 104–5; *see also* zippers
co-design process 2, *10*, 32–4, 53, 92, 97–113, 119–36, 141–64, 168–91
Cocjin, Sam 173–8
computerized knitting 156, *157*
constraints, design 49, 58, 60, 61
construction techniques 73–5

Cookman, Helen 3–5, *4*, 30
Cooper, Pamela 168–72
Coughlin, April 158–64
Cragg, Nina *34*, *35*
Crowther, Kaitlin 158–64

Desai, Uma 148–52
Design Requirements method 58–68
Dewey, Chamiah 72–3, *73*
disabled designers 5, 30, 34, 96, 195–6
Domsitz Jabara, Angela 114–18
draping 73, 74–5
dress forms *74*, 75
dresses, wrap *5*, 30, 101–7
dressing 11–13, 56, 94, 137–40, 155, 186: aids and tools 20–1, *21*, 99; independent 16–17, 97–100, 101–7, 134, 138, 179–84; interdependent 18–20, *19*, 141–7; materials for ease of 71

Ease 148–52
education 199, 201–2, 203
elbow patches 80–1, *80*, 161, *161*
evaluation, design 87–8
eyewear 41–2, *41*

face mask, heated 173–8
fashion design 31–2, 200–3
FEA (Function, Environment, and Aesthetics) model 36, 37–9
FEELDOM 13, *13*
footwear *12*, 197–8, *198*
Fowles, Ellen 11, *11*, 74, *74*
functional fashion 4–5, *4*, *5*, *8*, 37

Gaskin, Abby 9–11, *10*, 43, *43*
Gaui, Renata 158–64
Gaussite, Roxinne 101–7
Genentech 195, *195*
Gilinsky, Josh 38, *38*
glove guards 162–4, *163*, *164*
Glover, Chrissy 148–52
Guan, Kailu 125–32
Guo, Yiyun 108–13
gussets *5*, 78–9, *78*, *79*, 152

HAAT (Human Activity Assistive Technology) model 3, 36, 40–2

Handler, Rachel 119–24
Haren, Juliette S. Van 119–24
Heartist 8, *8*, 14–15, *14*, *15*, 16
heated jacket 185–91
Herman, Jerron 34, *34*
Hermès 27, *28*
human factors 44, 137–40
Hwang, Tiffany 19–20, *19*

inclusive design 13, 30–1, 48, 53, 92, 94–6
interactive garments 40, 51, 165–7, 179–84
interdisciplinary design *26*, 36, 37, 48, 167

jackets 15, *15*, *16*, 116, *117*, 153–7: Access & Closure 51–72, *87*; *see also* outerwear
JanSport 13
Jones, Dorothy 51–72, *87*
Joshpack 38, *38*
journey maps 37, 56, 111, *111*, 121, *122*, 170
Jun, Grace, Access & Closure 51–72, *87*
Jun, Jin Ah 20, *20*
June Adaptive 11, *12*

Kani, Arash 173–8
Kawakubo, Rei 28, *29*
Kim, Heeyoung 179–84
Kohl's Adaptive Clothing 197

Ladau, Emily 53, 153–7
Levi Strauss & Co 3–4, *4*, *6*
Liao, Julia *21*, 97–100, 133–6
Little Black Bag 108–13
LIULID 179–84
loops 33, *33*, *69*, 75–6, *75*, *76*
Lourie, Teresa 34, *35*
Luo, Ray 101–7

magnets 73, *73*, 77–8, *77*, *78*, 143: attaching to wheelchairs 104–5
Mahoney, Maggie 158–64
Mallon, Christina 21, *21*, 94–6, 97–100, 133–6
materials 32, 71, 114–18: abrasion resistant 161, *161*, 162–3, *163*;

breathable 106, 116, 129, 131, 160, 161, 185; conductive 51, *52*, 188; interactive 165–7, 183; performance 9, 112, 117, 118, 142, 146, 160; performance testing 115, 131, *131*, 161; researching 70, 106, *106*, 130–1, 150, 161, 162, *162*, 163, 172, 175, 181, *181*, 188, *188*; for sensory preferences and durability 149, 150–1, *151*; stretch 15, 116, *143*, 144, 151, 152, 156–7, *157*; sustainable 9, *10*, 11, 73; tactile textures 39, *39*, *40*; water-resistant 11, 112, 127, 131, 146, 160, 161, *188*; wool 100, 115, 116, 117, *117*, 123, *123*, 182; wrinkle-resistant 27, 115–16
Mehle, Silvo 185–91
Merrell, Ruthie 141–7
method 42–4
Midi-Rox 101–7
models 36–42
Modiste 153–7
MOHO (Model of Human Occupation) 36, 39–40, 127
Moy, Justin 141–7
Mury, Aimee and Eliza 148–52

NOT *18*, 19

objectives, project 50, 53–4, *55*
occupational therapists 20, 99, 101, 102, 137–40
Oh, Yangkyoon 153–7
Okunishi, Jensin 185–91
Open Style Lab 43, *43*, 75, 94, 95, 126, 195
origami 181–2, *181*
Osipow, Julie 179–84
outerwear 32–3, *33*, 40, *40*, 73, *73*: independent dressing 97–100; wheelchair users 125–32, 141–7, 158–64, 185–91

Paganelli, Nicholas 153–7
Paget, Mia 173–8
pain points 56, *57*, 172, *172*, 175
pants: catheter pocket *17*; closures 8, 82–3, *83*; front pockets 82, *82*; functional fashion 3–4, *4*, *6*; loops 76, *76*; for seated body 14–15, *15*, 83–5, *84*, *85*, 179–84; for wider access 9, *9*
Parikh, Priyal 141–7
Patti and Ricky 198
Peng, Fanyun 168–72

performative garments 34, *34*
plastic optical fiber 166, 183
pleats *7*, 27, 86, *86*, 181, *181*, 182: action *64*, 70, *70*, 79–80, *79*
pockets 27, 82, *82*, 183, 196, *196*
Poh, Claudia 16–17, *17*, *21*, 97–100, 133
principles, design 30–2
problem, defining 64
prosthetics 3, 5, *9*, 119–24
prototyping, iterative 68–72
pulley system 182–3, *183*

qXgo 125–32

range of motion 51, *51*, *52*, 70, *98*, 99, 145, 160
Rebirth Garments 196–7, *197*
research: disability 49; observational 50, 53–4, *55*, 56
Revolve 158–64
ribbing neckbands 81, *81*
Riley, Elizabeth 148–52
Roche, Colleen 108–13
Rudek, Nya 34, *35*

Samsethsiri, Prow 141–7
Sapala, Andrew J. 168–72
seams 150, *151*, 161
Sehringer, Larissa 40, *41*
shirts 5, *11*, 39, *39*, *155*: closures 8, 77–8, *77*, *78*; gussets 78–9, *78*, *79*
skating clothes 9–11, *10*, 43, *43*
skin sensitivity 13, 71, 148, 196: Ease 148–52
skort 119–24
sleeves: loops on *69*, 75–6, *75*, *76*; raglan 69, *69*, 128, *128*, *149*, 152; for seated body 85–6, *85*, *86*; twisted drop 74, *74*; Slick Chick 27–8
social integration 14–16, 42, 154
Social Model of Disability 36, 42
Social Surge 195–6, *196*
socks *12*, 20
SOUR 74, 75
spinal support 168–72
Stewart, Ada 179–84
stigma, disability 42, 108, 174
suit, tailored 153–7
swimwear 5, 194, *194*
Swipe 133–6

T-shirts 148–52
Tan, Jeanne 165–7
Target Adaptive 13

techniques 49, 72–86
terminology, using preferred 53
thermoplastic polyurethane 182, 184
3D printing 39, 40, 130, *130*, 133–6, 182, *182*
Tommy Adaptive 8–9
Tranquilli, Michael 101–7, 137–40
Trans-skirt 119–24
Tupkary, Kalyani 108–13
Two Blind Brothers 13

UCD (user-centered design) 42–3
UCQC 32–3, *33*
underwear 3, 28
universal design 7, 30–1, 137, 195–6
Unparalleled *21*, 97–100
User Persona template 54, *55*

vacuum forming process 175–6, *176*, 177, 178
Versa Vest 168–72
visibility: clothing 130–1, 146, 161, *161*; increasing disability 194–7, 199

Wang, Eraince 119–24
Wardrop, Alyssa 101–7
Warmed Bomber 185–91
Werable 16–17, *17*
wheelchair users: bags 13, *13*, 108–13; outerwear 125–32, 141–7, 158–64, 185–91; pants 14–15, *15*, 83–5, *84*, *85*, 179–84; sleeve designs 85–6, *85*, *86*; wrap dress 101–7; workplace attire 14–15, *14*, *15*, *16*, 42, 98, 153–7
Wu, Grace 179–84

Xiao, Ying 168–72

Yuan, Célestine *37*, 38, *38*

Zappos Adaptive 197–8, *198*
Zhang, Nuomeng 119–24
Zhang, Yuchen 141–7
Zhang, Yue 108–13
Zhao, Chengcheng 125–32
ZipBack Jacket 141–7
zippers *12*, 72, 196, *196*, *198*: bag 40, *41*, 111–12, *111*, *113*; loops along 33, *33*; pant 9, *9*, 82–3, *83*, *84*; sleeves 69, *69*, 74
Zorec, Petja 185–91